Imaging Plasma Density Structures in the Soft X-Rays Generated by Solar Wind Charge Exchange with Neutrals

David G. Sibeck · Soft X-Ray Imaging International
Team

Imaging Plasma Density Structures in the Soft X-Rays Generated by Solar Wind Charge Exchange with Neutrals

Previously published in *Space Science Reviews* Volume 214,
Issue 4, Article 79, 2018

 Springer

David G. Sibeck
Goddard Space Flight Center
Greenbelt, MD, USA

Soft X-Ray Imaging International Team

ISBN 978-94-024-1691-6

Front cover image: Simulated soft X-ray image of the Earth's magnetosheath and northern cusp from an equatorial post-noon vantage point. Each pixel results from a line-of-sight integration over emissions proportional to the product of neutral exospheric densities and the flux of high charge state solar wind ions. Credit: Kip Kuntz.
Back cover image: ROSAT All Sky Survey map of the ¼ keV sky before the removal of the streaks, named Long Term Enhancements (LTEs), which result from the soft X-rays emitted when high charge state solar wind ions exchange electrons with exospheric neutrals in the immediate vicinity of the Earth. The map is in galactic coordinates with (l,b) = (0°, 0°) at the center and longitude increasing to the left. The diffuse X-ray surface brightness increases from purple to white [Plate 38 of Snowden et al., 1995].

Printed on acid-free paper

This Springer imprint is published by the registered company Springer Nature B.V.
The registered company address is: Van Godewijckstraat 30, 3311 GX Dordrecht, The Netherlands

Contents

About the Authors

David Sibeck

David G. Sibeck is a Heliophysicist in NASA/GSFC's Space Weather Laboratory specializing in studies of the solar wind-magnetosphere interaction with an emphasis on terrestrial and planetary magnetosheaths and their inner magnetopause and outer bow shock boundaries. He received the American Geophysical Union's (AGU) Macelwane award in 1992 and went on to serve as Chair of the National Science Foundation's Geospace Environment Modeling program and as president of the AGU's Space Physics and Aeronomy section.

Robert Allen

Robert C. Allen is a post-doctoral fellow in the Space Physics Group at the Johns Hopkins Applied Physics Laboratory. He received his PhD in Physics from the University of Texas at San Antonio/Southwest Research Institute joint program in 2017, and his research interests lie in collisionless plasma phenomena. This includes sources and evolution of magnetospheric plasma, wave-particle interactions, and space flight instrumentation. His experience includes comprehensive studies of solar wind-originating plasma injection and subsequent evolution in the magnetospheres of Earth and Saturn, investigations on EMIC wave generation and propagation in the Earth's magnetosphere, as well as space flight instrument calibration.

Homayon Aryan

Dr. Homayon Aryan graduated in 2011 from the University of Sheffield with a class I master's degree in Aerospace Engineering with a private pilot license. He pursued his interest in space research and obtained a PhD in 2015 in space physics at the University of Sheffield. He won a prestigious NASA postdoctoral Fellowship award to expand his research at NASA Goddard Space Flight Center from 2015 to 2018. Dr. Aryan's research is mainly focused on improving plasma wave models that will ultimately increase the accuracy of radiation belt modeling.

Dennis Bodewits

Dennis Bodewits, currently at Auburn University, works on the activity and evolution of comets using both telescopic observations and in-situ exploration with planetary missions such as Deep Impact and Rosetta.

Pontus C. Brandt

Since 2000, Dr. Brandt has been employed by The Johns Hopkins University Applied Physics Laboratory, Maryland, where he has served as the Assistant Supervisor of the Space Physics Group, the Principal Investigator for two instruments on the JUpiter ICy moons Explorer mission managed by the European Space Agency, and the Project Scientist on a NASA-funded study of a pragmatic Interstellar Probe. Dr. Brandt received his PhD in space plasma physics from the Swedish Institute of Space Physics in Kiruna, Sweden in December, 1999. He has worked on several space missions including the Swedish microsatellite Astrid, the NASA IMAGE, TWINS, Van Allen Probes and Cassini missions, and the ESA Mars and Venus Express missions.

Graziella
Branduardi-Raymont

Graziella has been fascinated by astronomy and space research since she was a teenager. After receiving a degree in Physics at the University of Milano, Italy, and a PhD in X-ray Astronomy at University College London (UCL), she worked at the Harvard-Smithsonian Center for Astrophysics, USA, and then returned to UCL Mullard Space Science Laboratory where she is based and is Professor of Space Astronomy. Graziella has participated in major X-ray observatory missions over many years: Copernicus, Ariel 5 and the Einstein Observatory in the 1970s, EXOSAT in the 80s, ROSAT in the 90s. She is Co-Investigator for the Reflection Grating Spectrometer (RGS) operating on board XMM-Newton since 1999, and was project manager for the MSSL hardware contribution to the RGS. She is now co-leader of the SMILE mission with Prof. Chi Wang (CAS/NSSC).

Greg Brown

Greg Brown is a physicist in the LLNL Physics Division's Radiative Properties group. Greg began his work at LLNL in 1993 as a graduate student in the spectroscopy group at the Electron Beam Ion Trap facility, where he studied X-ray emission from Fe L-shell ions. After being awarded his PhD from Auburn University in 2000, Greg worked as a post-doc in the microcalorimeter group at NASA's Goddard Space Flight Center. While at Goddard, Greg became the co-lead scientist for calibration of the X-ray microcalorimeter, known as the XRS, currently on the Suzaku X-ray Observatory. In the fall of 2004 Greg returned to LLNL, where his research centered on experiments using the Electron Beam Ion Trap (EBIT). Greg's interests include high resolution spectroscopy for benchmarking atomic spectral models, laboratory simulation of plasmas found in celestial sources, X-ray emission from comets, planetary atmospheres and stellar coronae, and also the development and calibration of spectrometers for space flight and for diagnosing high energy density plasmas. Greg is a guest observer for the Swift Gamma Ray observatory and the Suzaku X-ray observatory.

Jennifer Carter

With a background in X-ray astronomy and a focus on exospheric and cometary X-rays, using data from the XMM-Newton and Swift missions, Jenny studies large-scale solar wind-magnetosphere-ionospheric coupling. Her research interests include the relationship between field-aligned acceleration and various auroral emission features observed in the ionosphere, and phenomena such as transpolar auroral arcs.

YUireska Collado-Vega

Dr. Yaireska (Yari) Collado-Vega is originally from Puerto Rico. Her research focuses on solar wind-magnetosphere interaction, instabilities, and transient events. It includes data analysis of multi-spacecraft observations from different missions and analysis of Magnetohydrodynamic (MHD) simulations. She is also interested in dayside-magnetopause magnetic reconnection, magnetopause standoff position changes, and soft X-ray imaging of the magnetosphere where she is part of new project missions. She is a member of the Community Coordinated Modeling Center (CCMC) team, is a senior space weather forecaster and the lead of the space weather forecaster team. She conducts education and public outreach for the CCMC team and the Heliophysics Science Division including social media events and live tv/radio interviews. She is the space weather subject matter expert and CCMC representative on the NASA Space Science Education Consortium and has been a Co-Investigator on several funded Science Innovation Fund proposals on a data mining tool development for vortices on the Earth's magnetopause.

Michael R. Collier

Michael R. Collier is currently a civil servant at NASA/GSFC. He has fabricated, calibrated, commanded and analyzed data from many flight hardware projects over about 30 years in the field. He has launched five instruments into space as hardware Principal Investigator, three low energy neutral atom imagers and two soft X-ray imagers. He is the author or coauthor of over 100 peer-reviewed scientific articles covering solar wind, heliospheric, terrestrial magnetospheric and outer planets physics.

Hyunju K. Connor

Hyunju K. Connor is a space scientist who models solar wind interaction with the Earth's Magnetosphere–Ionosphere–Thermosphere (MIT) coupled system. She has studied magnetic reconnection, particle acceleration near the magnetopause, and MIT coupling processes. She has also supported the Solar wind–Magnetosphere–Ionosphere Link Explorer (SMILE) mission by simulating soft X-ray images and developing the imaging processing techniques.

Thomas Cravens

T.E. Cravens is a professor of Physics and Astronomy at the University of Kansas and a Fellow of the American Geophysical Union. His research has emphasized how solar radiation and the solar wind interact with planets and other objects in our solar system, such as comets. He is particularly interested in sources of X-ray emission throughout the solar system, but is also involved in a number of studies associated with the Saturn system, using Cassini data, and with Mars, using MAVEN data.

Yuichiro Ezoe

Yuichiro Ezoe is an associate professor at Tokyo Metropolitan University. He received his BS, MS, and PhD degrees in physics from the University of Tokyo in 1999, 2001, and 2004, respectively. His current research interests include X-ray optics, X-ray microcalorimeters and astronomical objects including solar system objects, exoplanets, young stars, and blackholes.

Mei-Ching Fok

Dr. Mei-Ching Fok's main research interests are studies of the radiation belts and ring current during geomagnetic active periods, understanding the mechanisms responsible for their intensification and decay by numerical modeling and data analysis. She has developed a complex model, called the Comprehensive Inner Magnetosphere-Ionosphere (CIMI) Model, which computes and predicts energetic plasma fluxes in the radiation belts/ring current region.

Massimiliano Galeazzi

Massimiliano Galeazzi is a professor of physics at the University of Miami working on the properties of the Diffuse X-ray Background. He is the PI of the Diffuse X-rays from the Local Galaxy (DXL) sounding rocket mission to study the properties of the Local Hot Bubble and Solar Wind Charge eXchange.

Olga Gutynska

Olga Gutynska received her PhD in Prague, Charles University in 2011 with the main scientific focus on magnetosheath turbulent processes. After receiving her PhD, she took a postdoctoral position at CNRS in Toulouse, France, where she studied mirror mode waves in Saturn's magnetosheath, and in 2012 she was awarded a postdoctoral position at NASA, Goddard, USA. She has since widened her interests by studying waves in the foreshock region and substorm-related processes in the magnetosphere. During her stay at GSFC she also assisted the team developing a novel wide field-of-view soft X-ray imager. At present, she is back at Charles University, Prague, where she is continuing her research in the aforementioned areas.

Mats Holmstrom

Mats Holmstrom is a scientist at the Swedish Institute of Space Physics in Kiruna, Sweden. He received his PhD from Uppsala University, Sweden, in 1997. His main interest is computer modeling of space plasmas using particle methods. He is Principal Investigator of the ASPERA-3 instrument on Mars Express.

Syau-Yun Hsieh is a physicist working at the Johns Hopkins University Applied Physics Laboratory (JHU/APL) in Laurel, Maryland. Her research interests include space- and ground-based imaging, and space weather models and applications. She leads ground-based imaging development in the Space Exploration Sector at JHU/APL and has participated in many space imaging missions including DART/DRACO, DMSP/SSUSI, TIMED/GUVI, and IMAGE/HENA.

Syau-Yun Hsieh

Kumi Ishikawa is an aerospace project research associate at the Institute of Space and Astronautical Science (ISAS)/Japan Aerospace Exploration Agency (JAXA). She earned her MS and PhD degrees in physics from Tokyo Metropolitan University in 2010 and 2013, respectively. Her current research interests are X-ray optics, X-ray microcalorimeters, and observations of solar system objects and exoplanets.

Kumi Ishikawa

Dimitra Koutroumpa is currently a researcher of the CNRS - Centre National de Recherche Scientifique, working at LATMOS—Laboratoire Atmosphères, Milieux, Observations Spatiales laboratory. LATMOS is affiliated with two universities: Université de Versailles Saint-Quentin-en-Yvelines (member of the Université Paris-Saclay federation) and Université Pierre et Marie Curie (UPMC, member of the Sorbonne Universités federation). She received her PhD in Astrophysics in 2007 from the Pierre-and-Marie-Curie University in Paris. Her research interests include X-ray and UV gas emission and charge-exchange phenomena in the heliosphere with application to various astrophysical systems.

Dimitra Koutroumpa

Kip Kuntz

K.D. Kuntz is a Principal Research Scientist in the Rowland Department of Physics and Astronomy at the Johns Hopkins University. His primary research interest has been the diffuse X-ray background due to the Milky Way's interstellar medium and halo, as well as the cosmic background. For the last decade, he has been working to characterize the X-ray foregrounds due to solar wind charge exchange in order to isolate the astrophysical signal.

Maurice Leutenegger

Maurice Leutenegger currently works in the Sciences and Exploration Directorate of NASA's Goddard Space Flight Center as a Research Scientist in the X-ray Astrophysics Division. He received his PhD in physics from Columbia University and is mainly interested in astrophysical X-ray spectroscopy and associated fields, including X-ray optics, detectors, atomic physics, highly charged ions, and plasma physics. He has worked on Astro-H, a joint mission of JAXA and NASA, which included an X-ray calorimeter imaging spectrometer. He also worked on laboratory astrophysics experiments using an X-ray calorimeter at the Lawrence Livermore Electron Beam Ion Trap facility.

Yoshizumi Miyoshi

Yoshizumi Miyoshi is a professor at the Institute for Space-Earth Environment Research at Nagoya University. He has studied geospace dynamics using integrated analyses with the satellite observations and ground-based observations as well as computer simulations. He is the project scientist of Japanese ERG (Arase) satellite mission. He has served as an editor of Annales Geophysicae of the European Geoscience Union and as an associate editor of the Journal of Geophysical Research of the American Geophysical Union.

Frederick Scott Porter

Scott Porter received his PhD in Physics from Brown University in 1993. His dissertation work was on developing a prototype superfluid helium, solar neutrino telescope. He has since worked on a number of X-ray spectroscopy instruments first as a postdoctoral fellow at the Naval Research Laboratory, then joining the X-ray Astrophysics Laboratory at NASA/GSFC in 1995. He has flown 12 sounding rockets and three orbiting observatories based on cryogenic X-ray spectrometers. He currently leads the high energy laboratory astrophysics program at GSFC, is an instrument scientist on the XRISM/Resolve instrument, a Co-I on the CuPID CubeSat, and is a Co-I on the Athena/X-IFU instrument. His astrophysical interests center around superbubbles, the local interstellar medium, and simulating X-ray emitting plasmas in the laboratory.

Michael Purucker

Michael Purucker is a planetary scientist and geophysicist who investigates the processes that produce magnetic fields in the cosmos. He is the head of the Planetary Magnetospheres Laboratory at Goddard Space Flight Center and the president-elect of the Geological Society of Washington.

Andrew Read

Andrew Read is an X-ray astrophysicist working on ROSAT, XMM-Newton, Chandra, Swift, and Astrosat. He is a Co-I on the XMM-Newton Slew Survey and Manager of the XMM-Newton Background Working Group. He is heavily involved in the ESA-CAS SMILE mission.

Joachim Raeder

Joachim (Jimmy) Raeder is a professor of physics at the University of New Hampshire with a joint appointment in the UNH Space Science Center. He received his PhD from the Universität zu Köln (Germany) in geophysics and applied mathematics. He is best known as the original developer of the Open Geospace General Circulation Model (OpenGGCM), which is a coupled numerical model of Earth's magnetosphere, ionosphere, and thermosphere that is open for use by the scientific community at NASA's Community Coordinated Modeling Center. Dr. Raeder is a co-Investigator on NASA's THEMIS mission, and he has served on numerous NASA, NSF, and NRC panels, including a term as chair of the NSF Geospace Environment Modeling (GEM) program steering committee. Dr. Raeder's research focuses on many aspects of plasma processes in the Earth's magnetosphere, numerical modeling, and space weather.

Ina Robertson

Ina Piket Robertson obtained her PhD from the University of Kansas in 2001, having written a dissertation entitled "Heliospheric X-rays due to Solar Wind Charge Exchange." Her work explored the spatial and temporal variation of the emission due to both magnetospheric and heliospheric solar wind charge exchange, both in general and applied to the ROSAT All-Sky Survey. Her work has since extended to charge exchange in the lunar exosphere.

Andrei Samsonov

Andrey Samsonov received his PhD from St. Petersburg State University in Russia. His scientific interests include solar wind–magnetosphere interaction, magnetosheath and magnetopause, interplanetary shocks, and MHD modeling. Recently, he has studied the propagation of solar wind discontinuities through the magnetosheath and compared magnetopause models to observations. He will apply his knowledge and expertise in this area in to prepare for the forthcoming SMILE mission.

Steven Sembay

Steven Sembay specializes in the design and calibration of instrumentation for the detection of soft X-rays. He is the Principal Investigator of the European Photon Imaging Camera (EPIC) on ESA's XMM-Newton and the Principal Investigator of the Soft X-ray Imager (SXI) on the forthcoming ESA-CAS SMILE mission. An X-ray astronomer by training, Steve has become intrigued by the challenge of using global imaging in soft X-rays to study the interaction of the solar wind with the Earth.

Steven Snowden

Steve Snowden received his PhD from the University of Wisconsin-Madison in 1986, after which he spent five years as a post-doc at the Max Planck Institute for Extraterrestrial Physics working on the ROSAT project. Returning to the US, he worked on the ROSAT project at GSFC, later moving to the XMM-Newton project and ending up as the XMM-Newton Project Scientist. He has spent his professional career studying the cosmic soft X-ray diffuse background, primarily the emission coming from the Local Hot Bubble. The study of solar wind charge exchange X-ray emission was forced on him as it contributes a very significant background to his signal of interest.

Nicholas Thomas

Nick Thomas is a research physicist with a zeal for developing state-of-the-art instrumentation for X-ray astrophysical observations. Nick's interest in physics began as a student at John Glenn High School in New Concord, Ohio, and was further fostered as an undergraduate at the University of Pennsylvania. Both at Penn and as a graduate student at the University of Miami, Nick began his work in instrumentation with his primarily focus being on high altitude balloon-based observations. Following graduation, Nick accomplished a lifelong goal in joining NASA at Goddard Space Flight Center via contracting through the University of Maryland, Baltimore County with his work focusing on X-ray telescope instrumentation. His current work includes the DXL (Diffuse X-ray emission from the Local galaxy) sounding rocket program and the characterization of slumped micropore (a.k.a. "Lobster-eye") X-ray optics for STORM (Solar-Terrestrial Observer of Reconnection in the Magnetosphere) and ISS-TAO (Transient Astrophysics Observatory on the ISS).

Rudolf von Steiger holds a PhD from the University of Bern, where he also obtained habilitation. He worked as a scientist at various institutions including the University of Michigan, the University of Maryland, the University of Bern, and the International Space Science Institute (ISSI). Today, he is Director of ISSI in Bern, Switzerland and Professor at the University of Bern. His scientific interests include the composition, charge state distributions, and other properties of heavy ions in the solar wind.

Rudolf von Steiger

Professor Walsh is an experimental space scientist who focuses on the coupling of energy from the solar wind into the Earth's magnetosphere and ionosphere. His research utilizes measurements from various large-scale NASA and ESA space missions, and he also develops small satellite constellations.

Brian Walsh

Simon Wing is presently a Principal Professional Staff Physicist at The Johns Hopkins University Applied Physics Laboratory and an Adjunct Associate Professor at University of Maryland University College. His research interests include modeling the open field line particle precipitation; modeling the magnetotail pressure, density, and temperature; geosynchronous environment; solar wind-magnetosphere interaction; and space weather.

Simon Wing

Foreword

Imaging the Sun-Earth connection is a goal that has been pursued since the first in-situ observations of the outer boundaries of the Earth's magnetic field in the early 1960s. Although auroras were observed much earlier from the ground (the first written report on aurora in China was in 2600 B.C.), their basic understanding and their link with the Sun was only fairly recently achieved. Since then, many spacecraft have explored the ionosphere, magnetosphere and solar wind to measure particles and fields in situ and quantify the solar wind mass, energy and momentum being transferred to the Earth. Although in-situ measurements provide the ground truth on physical processes at work in the Sun-Earth connection, they do not provide a global view of the Earth's magnetic field and its interaction with the solar wind. On the other hand, such global scales can be obtained with new state-of-the-art instruments that make large-scale images of the magnetosphere.

"Imaging of auroral emissions from earth satellites is an extremely useful tool in the analyses of observations of the complex plasma phenomena which are encountered at low altitude over the earth's ionosphere and at the greater distances within the magnetosphere," noted L.A. Frank while developing the first global imager of the aurora to be flown on board Dynamics Explorer 1. The global images of the full auroral oval have revolutionized our understanding of the magnetosphere and in particular geomagnetic storms and substorms. 20 years later, NASA launched the Imager for Magnetopause-to-Aurora Global Exploration (IMAGE) spacecraft that imaged not only the aurora but also the plasmasphere and the ring current. Less than ten years later, the Interstellar Boundary Explorer (IBEX), whose science objective was primarily to map the boundary between the Solar System and interstellar space, obtained the first images of the magnetotail, magnetopause and polar cusp through energetic neutral atom measurements. However, such images required between 11 and 20 hours of acquisition time and could not be used to monitor the solar wind-magnetosphere interaction, which varies on a time scale of minutes. Such temporal and wide fields-of-view can be obtained with the new generation of instruments collecting soft-X rays generated by solar wind charge exchange.

This book provides an extensive review of the solar wind interaction with not only Earth but also Mars and Venus, the Moon and Comets, and the interstellar medium. It describes the charge exchange process between highly ionized solar wind ions and hydrogen atoms from the geocorona, which produces soft X-ray emissions. Such soft X-ray emissions can then be used to image large portions of the solar wind's interaction with planetary obstacles and track the dayside plasma boundaries generated by these interactions. Based on knowledge acquired in X-ray astronomy, wide field-of-view and lightweight X-ray cameras are being developed; prototypes have been flown on sounding rockets, and the first instruments will fly in the next few years. Global simulations of the Earth, Mars and Venus and the heliosphere are essential to identify the best targets for soft X-ray observations and to verify that the science objectives—in particular spatial and temporal resolutions—of future instruments can be achieved successfully.

This book reviews all the basic elements needed to image structures with soft X-rays. In particular, the development of soft X-ray imaging techniques will reveal the "invisible" terrestrial magnetic field. It will certainly be used as a textbook by scientists and engineers involved in the implementation of the new ESA-China SMILE mission, to be launched in a few years' time.

C. Philippe Escoubet
ESA, ESTEC, The Netherlands

Space Sci Rev (2018) 214:79
https://doi.org/10.1007/s11214-018-0504-7

Imaging Plasma Density Structures in the Soft X-Rays Generated by Solar Wind Charge Exchange with Neutrals

David G. Sibeck[1] · R. Allen[2] · H. Aryan[3] · D. Bodewits[4] · P. Brandt[2] ·
G. Branduardi-Raymont[5] · G. Brown[6] · J.A. Carter[7] · Y.M. Collado-Vega[3] ·
M.R. Collier[3] · H.K. Connor[8] · T.E. Cravens[9] · Y. Ezoe[10] · M.-C. Fok[3] · M. Galeazzi[11] ·
O. Gutynska[3] · M. Holmström[12] · S.-Y. Hsieh[2] · K. Ishikawa[13] · D. Koutroumpa[14] ·
K.D. Kuntz[15] · M. Leutenegger[3,16] · Y. Miyoshi[17] · F.S. Porter[3] · M.E. Purucker[3] ·
A.M. Read[7] · J. Raeder[18] · I.P. Robertson[9] · A.A. Samsonov[19] · S. Sembay[7] ·
S.L. Snowden[3] · N.E. Thomas[3,16] · R. von Steiger[20] · B.M. Walsh[21] · S. Wing[2]

Received: 7 December 2017 / Accepted: 3 April 2018 / Published online: 12 June 2018
© The Author(s) 2018

Abstract Both heliophysics and planetary physics seek to understand the complex nature of
the solar wind's interaction with solar system obstacles like Earth's magnetosphere, the ionospheres of Venus and Mars, and comets. Studies with this objective are frequently conducted
with the help of single or multipoint *in situ* electromagnetic field and particle observations,
guided by the predictions of both local and global numerical simulations, and placed in con-

✉ K.D. Kuntz
  kkuntz1@jhu.edu

  D.G. Sibeck
  David.G.Sibeck@nasa.gov

[1] Code 674, NASA/GSFC, Greenbelt, MD, USA

[2] The Johns Hopkins University Applied Research Laboratory, Laural, MD, USA

[3] NASA/Goddard Space Flight Center, Greenbelt, MD, USA

[4] University of Maryland College Park, College Park, MD, USA

[5] Mullard Space Science Laboratory, University College London, London, UK

[6] Lawrence Livermore National Laboratory, Livermore, CA, USA

[7] University of Leicester, Leicester, UK

[8] University of Alaska, Fairbanks, AK, USA

[9] University of Kansas, Lawrence, KS, USA

[10] Department of Physics, Tokyo Metropolitan University, Tokyo, Japan

[11] University of Miami, Miami, FL, USA

[12] Swedish Institute of Space Physics, Kiruna, Sweden

[13] JAXA/Institute of Space and Astronautical Science, Sagamihara, Kanagawa, Japan

[14] LATMOS/IPSL, UVSQ – Université Paris-Saclay, UPMC – Université Paris 06, CNRS,
  Guyancourt, France

[15] Johns Hopkins University, Baltimore, MD, USA

text by observations from far and extreme ultraviolet (FUV, EUV), hard X-ray, and energetic neutral atom imagers (ENA). Each proposed interaction mechanism (e.g., steady or transient magnetic reconnection, local or global magnetic reconnection, ion pick-up, or the Kelvin-Helmholtz instability) generates diagnostic plasma density structures. The significance of each mechanism to the overall interaction (as measured in terms of atmospheric/ionospheric loss at comets, Venus, and Mars or global magnetospheric/ionospheric convection at Earth) remains to be determined but can be evaluated on the basis of how often the density signatures that it generates are observed as a function of solar wind conditions. This paper reviews efforts to image the diagnostic plasma density structures in the soft (low energy, 0.1–2.0 keV) X-rays produced when high charge state solar wind ions exchange electrons with the exospheric neutrals surrounding solar system obstacles.

The introduction notes that theory, local, and global simulations predict the characteristics of plasma boundaries such the bow shock and magnetopause (including location, density gradient, and motion) and regions such as the magnetosheath (including density and width) as a function of location, solar wind conditions, and the particular mechanism operating. In situ measurements confirm the existence of time- and spatial-dependent plasma density structures like the bow shock, magnetosheath, and magnetopause/ionopause at Venus, Mars, comets, and the Earth. However, in situ measurements rarely suffice to determine the global extent of these density structures or their global variation as a function of solar wind conditions, except in the form of empirical studies based on observations from many different times and solar wind conditions. Remote sensing observations provide global information about auroral ovals (FUV and hard X-ray), the terrestrial plasmasphere (EUV), and the terrestrial ring current (ENA). ENA instruments with low energy thresholds (~ 1 keV) have recently been used to obtain important information concerning the magnetosheaths of Venus, Mars, and the Earth. Recent technological developments make these magnetosheaths valuable potential targets for high-cadence wide-field-of-view soft X-ray imagers.

Section 2 describes proposed dayside interaction mechanisms, including reconnection, the Kelvin-Helmholtz instability, and other processes in greater detail with an emphasis on the plasma density structures that they generate. It focuses upon the questions that remain as yet unanswered, such as the significance of each proposed interaction mode, which can be determined from its occurrence pattern as a function of location and solar wind conditions. Section 3 outlines the physics underlying the charge exchange generation of soft X-rays. Section 4 lists the background sources (helium focusing cone, planetary, and cosmic) of soft X-rays from which the charge exchange emissions generated by solar wind exchange must be distinguished. With the help of simulations employing state-of-the-art magnetohydrodynamic models for the solar wind-magnetosphere interaction, models for Earth's exosphere, and knowledge concerning these background emissions, Sect. 5 demonstrates that boundaries and regions such as the bow shock, magnetosheath, magnetopause, and cusps can readily be identified in images of charge exchange emissions. Section 6 reviews observations by (generally narrow) field of view (FOV) astrophysical telescopes that confirm the presence of

[16] University of Maryland Baltimore County, Baltimore, MD, USA

[17] Nagoya University, Nagoya, Japan

[18] University of New Hampshire, Durham, NH, USA

[19] University of St. Petersburg, St. Petersburg, Russia

[20] International Science Space Institute, Bern, Switzerland

[21] Boston University, Boston, MA, USA

these emissions at the intensities predicted by the simulations. Section 7 describes the design of a notional wide FOV "lobster-eye" telescope capable of imaging the global interactions and shows how it might be used to extract information concerning the global interaction of the solar wind with solar system obstacles. The conclusion outlines prospects for missions employing such wide FOV imagers.

Keywords X-rays · Magnetosheath · Cusp · Instrumentation · Solar wind · X-ray background · Charge exchange · Comets · Planets

1 Introduction

Earth's magnetic field carves out a cavity in the oncoming solar wind known as the magnetosphere. Because the magnetosphere extracts all of the mass, momentum, and energy that powers geomagnetic storms from the solar wind, quantifying and understanding the flow of these quantities from the Sun outward through the heliosphere, through the Earth's magnetosphere, and into the Earth's ionosphere is one of the primary goals of the heliophysics discipline. Similar objectives govern the planetary discipline which seeks, amongst other tasks, to determine the nature of the solar wind's interaction with comets and the other planets within our solar system, and in particular to quantify the role that plasma processes play in the loss of their atmospheres. Once the conditions governing the occurrence patterns of the various fundamental processes (including reconnection, diffusion, instabilities, particle acceleration, and ion-neutral interactions) that control the mass, energy, and momentum flow are well understood, it will become possible to construct numerical simulations that provide accurate space weather predictions for the immediate environment of the Earth and other solar system objects (e.g., Bertucci et al. 2011).

Figure 1 presents results from state-of-the-art hybrid code simulations for the plasma interactions that occur in the vicinity of Venus, Mars, and the Earth. From a global perspective, the density structures, and thus the processes that govern these interactions exhibit many similarities. A bow shock (BS) separates the higher density magnetosheath plasma of solar wind origin from the solar wind itself. A sharp magnetopause or ionopause (I) separates the magnetosheath from the planetary obstacle, whether it be the high density ionospheres with plasmas of planetary origin at Venus and Mars or the low density magnetosphere at Earth. The panels for Venus and Mars show boundary locations for a stable interplanetary magnetic field (IMF) transverse to the Sun-planet line, while the third panel shows boundaries near Earth during the passage of a solar wind tangential discontinuity (TD).

Many micro- to macro-scale processes have been predicted and observed to occur in the vicinity of the bow shock and magnetopause, as well as throughout the foreshock, magnetosheath, and outer magnetosphere. These processes are often identified on the basis of the diagnostic density structures that they generate. Macroscale structures include the bow shocks, magnetosheaths, and either the ionopauses or magnetopauses that stand upstream from both comets and planets. The location and motion of these boundaries depend not only on the time-varying conditions within the solar wind but also on conditions within the magnetospheres and ionospheres. Mesoscale features include dawn/dusk asymmetries in foreshock and magnetosheath parameters, waves and riplets driven by variations in solar wind parameters or instabilities on the boundaries, boundary layers of intermingled magnetosheath and magnetospheric or ionospheric plasma, and cusps filled with magnetosheath-like plasma that link Earth's magnetopause to its ionosphere and atmosphere. Microscale features include the kinetic structures generated by wave-particle interactions within the

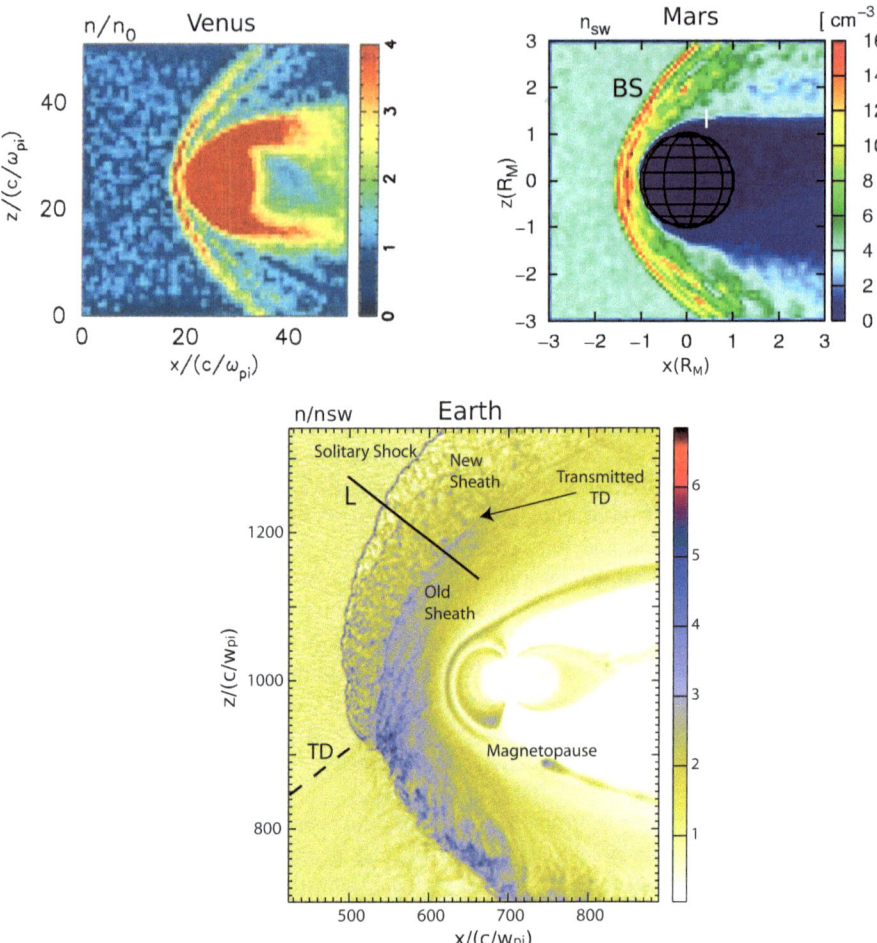

Fig. 1 Cuts through hybrid simulations of the solar wind-magnetosphere interaction showing the density of plasma with a solar wind origin at Venus (upper left panel, Bößwetter et al. 2007), Mars (upper right panel, Shimazu 2001b), and Earth (lower panel, Omidi and Sibeck 2007). The panels for Venus and Mars show boundary locations for a stable IMF transverse to the Sun-planet line, while the third panel shows boundaries near Earth during the passage of a solar wind TD at which the IMF rotates from northward and antisunward to dawnward. Here BS stands for bow shock and I for ionopause. Distances in the second panel are measured in planetary radii, in the first and third panel they are measured in terms of the ion skin depth $(c/\omega_{pi} = c[4\pi n_p e^2/M]^{-\frac{1}{2}} \sim 100$ km for $n = 10$ cm^{-3}, where c is the speed of light, M the mass of a proton, n_p the proton density, and e the charge of an electron). Densities in the first and third panel have been normalized to those in the solar wind. Note the multiple shock structures at Venus, the north/south asymmetries in bow shock and ionopause locations at Mars, and the complex shock structure at Earth

foreshock and the structure of the bow shock and density variations associated with magnetosheath waves.

The significance of each interaction process depends upon its spatial extent and the solar wind/magnetospheric conditions under which it occurs. While statistical studies of *in situ* observations can provide considerable information concerning the occurrence patterns of various phenomena, reconstructing the global configuration of density structures from isolated *in situ* measurements is no easy task. Global magnetohydrodynamic and, more recently, hybrid kinetic simulations provide considerable insight, but need validation by equally global measurements.

Pending the launch of constellation-type missions with thirty or more spacecraft in a wide array of orbits (e.g., The Magnetospheric Constellation, MC: Global Dynamics of the Structured Magnetotail, NASA 2004), imaging affords the best (and certainly the most cost-effective) means of (1) determining the overall configuration of the Earth's magnetosphere, (2) identifying the extent and significance of the processes governing the solar wind-magnetosphere interaction on the basis of their diagnostic plasma density signatures, and (3) validating the numerical simulations. Missions like DE-1 (Frank et al. 1981), Viking (Anger et al. 1987), Freja (Murphree et al. 1994), Polar (Frank et al. 1995; Imhof et al. 1995; Torr et al. 1995), and IMAGE (Mende et al. 2003) employed visible, ultraviolet, and X-ray imagers to take global pictures of the auroral oval, a region to which many of the most basic processes in the magnetosphere map. However, it can be difficult to determine both the nature of the processes and the locations of distinctive features in the magnetosphere that map to features in the auroral oval. The need for global images of the magnetosphere led to the launch of IMAGE and TWINS. These missions took extraordinarily fascinating and instructive images of the plasmasphere in extreme ultraviolet, of the cusp and subsolar magnetosheath in low-energy neutral atoms, of the auroral oval in previously unobserved far ultraviolet wavelengths, and of the ring current in higher energy neutral atoms. Discoveries included plasmaspheric shoulders and notches (Darrouzet et al. 2009), surprisingly slow plasmaspheric rotation (Burch et al. 2004), a hot oxygen geocorona (Wilson et al. 2003), and persistent proton auroras (Frey et al. 2003).

Observations of the global solar wind-magnetosphere interaction suitable for direct comparison with the predictions of global numerical models are now within reach. Operating from vantage points up to 49 R_E from Earth, the IBEX-Hi imager (Funsten et al. 2009) on the spinning ($\sim$ 4 rpm) Interstellar Boundary Explorer spacecraft (IBEX, McComas et al. 2009) has returned rastered images of the bow shock, magnetopause, and cusps in 0.9–1.5 keV energetic neutral atoms (ENAs), primarily hydrogen. The solar wind protons acquire electrons from exospheric hydrogen atoms and then proceed in their pre-exchange directions. Because the decelerated and thermalized solar wind protons gyrate around magnetosheath magnetic field lines, the pre-existing directions are effectively random over the expected scale lengths of magnetosheath phenomena, and the ENA flux seen in any direction is approximately proportional to the integrated line-of-sight (LOS) product of the plasma ion and exospheric neutral densities. Figure 2 presents examples for the magnetosheath (Fuselier et al. 2010) and cusp (Petrinec et al. 2011).

Strikingly different ENA flux levels are observed on LOS integrations that (1) remain solely in the low plasma and low neutral density solar wind, that (2) pass through the high plasma and moderate neutral density magnetosheath, that (3) pass through the high plasma and high neutral density cusps, and (4) that pass through the very low plasma and high neutral density equatorial or polar magnetosphere. Furthermore, the energies, composition, flux, and direction of the ENAs arriving at the observing location provide important information concerning the processes occurring at remote magnetospheric locations (Taguchi et al. 2004; Collier et al. 2005a; Hosokawa et al. 2008).

On the other hand, the $7° \times 7°$ single pixel IBEX-Hi imager requires times ranging from 11 to 20 hours to raster individual global ENA images, with inherent spatial resolutions in the noon-midnight meridional plane ranging from 3.7 R_E for spacecraft locations just outside the bow shock to 6.1 R_E at 49 R_E apogee. By contrast, cadences on the order of minutes to tens of minutes and spatial resolutions less than 1 R_E are needed to capture the

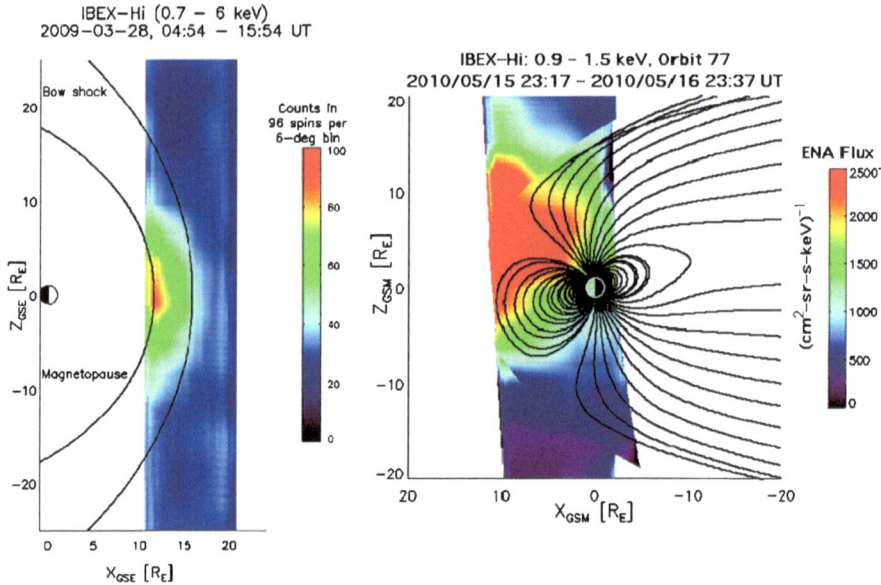

Fig. 2 ENA images of the dayside magnetosphere from the IBEX mission. The left panel presents measurements of ENAs from the subsolar magnetosheath (adapted from Fuselier et al. 2010), while the right panel shows ENAs from the cusps (Petrinec et al. 2011)

dynamics of the processes that govern the solar wind-magnetosphere interaction at the bow shock and magnetopause. Even if instantaneous global snapshots could be taken, the finite times-of-flight required for individual ENAs to arrive at the observing instrument would result in individual images representing the convolution of particles with different energies coming from different locations at different times.

An alternative method for imaging the magnetosphere offers the potential to obviate these problems. Exospheric neutral charge exchange with high charge state solar wind ions generates soft X-rays with energies from 0.05–2.0 keV. Currently existing wide field-of-view (FOV) soft X-ray telescopes provide an opportunity to image not only the dayside solar wind-terrestrial magnetosphere interaction, but also the interactions that occur at the Moon, Venus, Mars, and comets. This paper begins with a review of those scientific topics raised by modeling and past *in situ* missions that can be addressed by imaging missions. It then discusses the physical processes governing the generation of soft X-rays, in particular charge exchange with high charge state solar wind ions. Numerical simulations employ models for the solar wind composition, exosphere, solar wind-magnetosphere interaction, and soft X-ray background to predict the integrated LOS emission intensities observable by wide FOV soft X-ray imagers and define the cadence and spatial resolution required from such an imager. A review of previously reported observations by narrow FOV astrophysical telescopes demonstrates that the emissions are present at the predicted level from all of the proposed targets. Wide FOV soft X-ray telescopes capable of making global observations with the required spatial resolution and cadences have already flown and are scheduled for forthcoming missions. The features seen within the global images can be readily associated with density structures observed by *in situ* spacecraft on suitable orbits. The paper concludes with comments concerning prospects for wide FOV soft X-ray telescopes.

Fig. 3 Plasma structures generated by the solar wind's interaction with Earth's magnetosphere: solar wind (SW), bow shock (Bshock), and magnetopause (MP). Adapted from Wiltberger et al. (2015)

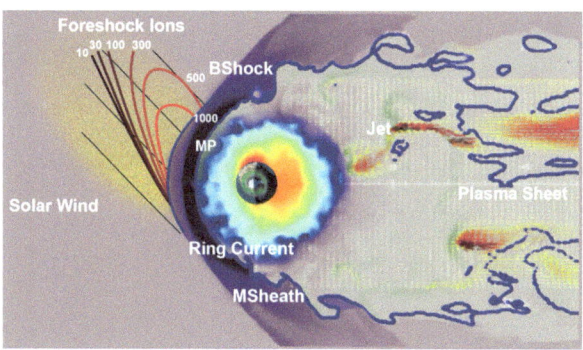

2 Scientific Objectives

Global images of the soft X-rays generated when high charge state solar wind ions (e.g., C^{6+}, O^{7+}, O^{8+}, Fe^{12+}) exchange charges with neutrals (e.g., H, H_2O) can provide crucial information concerning the nature of the solar wind's interaction with planetary atmospheres and magnetospheres, including those of the Earth, Venus, Mars, the Moon, and comets. As illustrated in Fig. 3, the reason for this is that the processes governing the interaction of the solar wind with these heliospheric obstacles generate a host of plasma density structures that can be used to diagnose the nature of those interactions. At the Earth, the size, shape, structure, and motion of the magnetopause and cusps provide important information concerning the global characteristics of magnetic reconnection, the strength of various magnetospheric current systems, and the response of the magnetosphere to varying solar wind and foreshock input. Observations of transients at the magnetopause and in the cusps quantify their extent and occurrence patterns, hence their significance to the overall interaction. Observations of the magnetosheath structure and its time variability provide the outer boundary conditions for the magnetosphere. The location of the bow shock yields information concerning the thermodynamics of the collisionless solar wind, while the structure of the bow shock defines its ability to reflect and energize particles, a fundamental heliospheric process. Observations of the foreshock are needed to understand and quantify the effects of the particles accelerated at the bow shock upon the bulk parameters of the incoming solar wind and therefore upon the overall solar wind-magnetosphere interaction.

There are parallel research problems to be addressed by imaging comets, the Moon, Venus, and Mars. These topics concern the interaction of the solar wind with obstacles that have little or no intrinsic magnetic field. In the cases of comets, Venus, and Mars, studies that focus on the location, structure, and motion of the bow shock and ionopause yield information concerning atmospheric loss rates. In the case of the Moon, studies focus upon the structure, composition, and sources of the tenuous lunar exosphere. This section describes potential research questions.

2.1 The Earth

We begin by considering those questions concerning the Earth's magnetopause, cusps, transients at the magnetopause and in the cusps, the magnetosheath, bow shock, and foreshock that can be diagnosed with the help of information concerning plasma density structures deduced from soft X-ray observations. We then address those questions concerning the processes that occur at comets, Venus, Mars, and the moon that can also be answered with the help of soft X-ray images.

Fig. 4 Results from an empirical model for the locations of the equatorial magnetopause as a function of (upper panel) 5 solar wind pressures (0.54–0.87, 0.87–1.47, 1.47–2.60, 2.60–4.90, and 4.90–9.90 nPa) and (lower panel) 6 values of IMF Bz (−6 to −4, −4 to −2, −2 to 0, 0 to 2, 2 to 4, and 4 to 6 nT) (Sibeck et al. 1991). The plots are in geocentric solar ecliptic (GSE) coordinates in Earth radii (R_E) with $R = \sqrt{y^2 + z^2}$

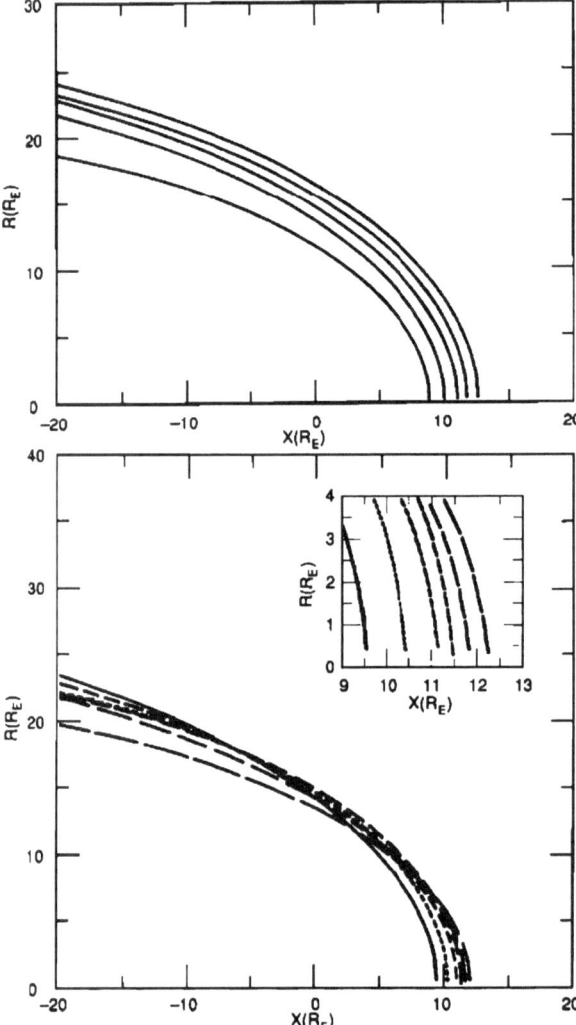

2.1.1 Earth's Magnetopause

A host of factors, including the solar wind thermal and dynamic pressures, the IMF latitude and cone angle, the dipole tilt, and the strength of various current systems within and bounding the magnetosphere determine the location of the magnetopause. Although they can predict widely divergent magnetopause locations for the same solar wind conditions (Samsonov et al. 2016), global magnetohydrodynamic simulations all indicate that the solar wind dynamic pressure and north/south component of the IMF are the most important factors determining magnetopause location (Lu et al. 2011). As illustrated in the top panel of Fig. 4, empirical studies based on large numbers of magnetopause crossings paired with time-averaged solar wind measurements suggest that the magnetopause expands and contracts in a self-similar manner in response to variations in the solar wind dynamic pressure (Sibeck et al. 1991; Roelof and Sibeck 1993; Lin et al. 2010; Wang et al. 2013). Some case studies disagree (Stüdemann et al. 1986). Both case studies (e.g., Kaufmann and Konradi

1969) and numerical simulations (Samsonov et al. 2015) confirm that the response of the magnetopause to step-function changes in the solar wind pressure is more complicated than self-similar contractions and expansions.

By contrast, in response to changes in the IMF orientation, the dayside magnetopause moves Earthward (Aubry et al. 1970), the cusps move equatorward (Newell et al. 1989), and the magnetotail flanks move outward (Maezawa 1975) during intervals of southward IMF orientation, thereby producing a blunter magnetosphere with a greater magnetopause flaring angle. This erosion, or inward motion of the dayside magnetopause and outward motion of the magnetotail magnetopause, can be attributed to magnetic reconnection, a process that removes magnetic flux from the dayside magnetosphere and adds it to the magnetotail, although it has recently been noted that a (small) portion of the inward motion may result from the enhancements of the pressure near the subsolar magnetosheath known to occur for the blunter magnetopause shapes during intervals of southward IMF orientation (Shue et al. 2013). Wiltberger et al. (2003) propose that magnetopause erosion results from (rather than causes) enhanced cross-tail currents. Soft X-ray images will provide an opportunity to determine the instantaneous shape of the global magnetopause and define its evolving response to solar wind variations, thereby distinguishing between the possibilities outlined above.

Because it is the dominant process enabling the transfer of solar wind mass, energy, and momentum into the magnetosphere, understanding reconnection is a fundamental heliophysics objective. In conjunction with solar wind observations, magnetopause locations and shapes can be used to deduce the magnetic field strengths just inside the magnetopause, the strengths of the relevant magnetospheric current systems, the amount of flux eroded from the dayside magnetosphere, and as a result, the global response of reconnection to varying solar wind conditions (Sibeck et al. 1991). At rest, the magnetopause lies along the locus of points where the sum of thermal and magnetic pressures balance in the magnetosheath and magnetosphere. Under both elastic and inelastic reflection hypotheses, the magnetosheath pressure applied locally to the dayside magnetopause is proportional to the fraction of the solar wind dynamic incident upon the flaring magnetopause surface (e.g., Spreiter et al. 1966). With the exception of the cusps, where plasma pressures are high, the total pressure applied by the magnetosheath to the magnetopause is balanced almost exclusively by the magnetic pressure just inside the magnetopause. However, the magnetic fields that contribute to this magnetic pressure are themselves just the sum of contributions from all magnetospheric current systems.

Thus, together with a measure of the solar wind dynamic pressure, soft X-ray observations of the location and shape of the magnetopause can be used to infer magnetic field strengths just inside the magnetopause and in turn variations in magnetospheric current systems as a function of solar wind conditions. In the case of reconnection, the relevant current systems are the Region 1 Birkeland current and, to a much lesser degree, the cross-tail current systems (Maltsev and Lyatsky 1975; Tsyganenko and Sibeck 1994). Operating in tandem, these current systems reduce dayside magnetospheric magnetic field strengths, transfer magnetic flux to the magnetotail, and allow the dayside magnetopause to move inward during intervals of southward IMF orientation. With their strengths inferred from observations of the dayside magnetopause location, the amount of flux eroded by reconnection from the dayside magnetosphere can be determined for any combination of solar wind or geomagnetic parameters (Sibeck et al. 1991; Shue et al. 2001).

Observations of magnetopause motion can be used to determine the time-dependence of reconnection. Although both *in situ* and ground-based observations provide evidence for steady and impulsive reconnection, the conditions governing when and where each occur

Fig. 5 Meridian scanning photometer measurements from Svalbard (adapted from Oksavik et al. 2005). The top panel presents the 630.0 nm line while the bottom panel is the 557.7 line. Periodic poleward moving enhancements are observed

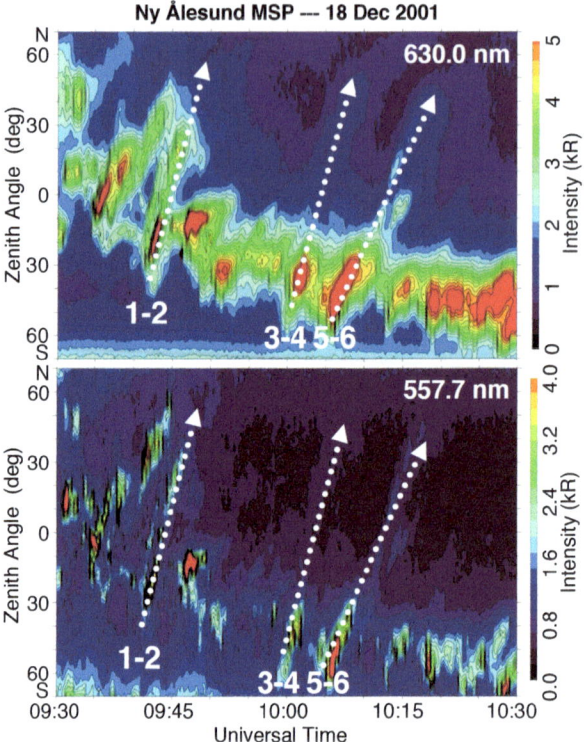

remain unknown. Drake et al. (2006) suggest that antiparallel magnetosheath and magneto-spheric magnetic fields favor steady reconnection along a single line, whereas shear angles less than 127° result in unsteady reconnection and the formation of magnetic islands or flux ropes. Steady reconnection predicts a gradual inward motion of the dayside magnetopause following southward IMF turnings, perhaps several Earth radii over a period of one to two hours (e.g., Aubry et al. 1970). It is not yet known how or whether the rate of this steady erosion changes with time. By contrast, sporadic reconnection models predict a sequence of abrupt earthward leaps, perhaps once each 8 minutes or so, corresponding to the equatorward jumps seen in ground-based radar and optical observations of the cusps when the IMF turns southward (Lockwood et al. 1989; Sandholt et al. 1998). Figure 5 shows one such sequence of events reported in ground-based observations of auroral emissions at 557.7 nm and 630.0 nm (Oksavik et al. 2005). With simultaneous solar wind observations, one can use soft X-ray observations to determine whether (Lockwood and Wild 1993) or not (Le et al. 1993) the bursts of reconnection corresponding to the inward magnetopause leaps are triggered by intrinsic magnetopause instabilities or fluctuations in the IMF orientation.

As a corollary, global images of the magnetopause location can be used to determine the time scale required for the magnetopause to move outward following a substorm onset or a northward IMF turning, and the mechanisms by which it does so. The outward motion of the magnetopause under these circumstances implies an addition of magnetic flux to the dayside magnetopause. The flux might be added by appending magnetosheath field lines to the dayside magnetosphere via either steady or unsteady simultaneous reconnection poleward of both cusps (Song and Russell 1992). Alternatively, the flux might be returned by sunward convection within the magnetosphere that continues even when the IMF turns northward (e.g., Øieroset et al. 1997). The rate of flux accretion remains unknown, but could be determined by tracking outward dayside magnetopause motion during intervals of northward IMF orientation.

Global perspectives can provide important information about the location and extent of reconnection. Component reconnection models predict erosion of the magnetopause along a tilted line passing through and centered on the subsolar point (Gonzalez and Mozer 1974; Laitinen et al. 2007). For many IMF orientations, antiparallel reconnection models predict reconnection at locations far from the subsolar point (Crooker 1979; Sandholt and Farrugia 2003). In the former case, magnetopause motion should begin at the subsolar point, in the latter case it should begin at locations away from the subsolar point. The manner in which reconnection spreads must also be determined. It could be initiated simultaneously over a wide region of the magnetopause as inferred from the sudden appearance of transient events in the high-latitude dayside auroral ionosphere (e.g., Lockwood et al. 1990), spread in the direction of the current at the speed of the current carriers for weak guide fields (Lapenta et al. 2006), or spread both along and opposite the current simultaneously at the Alfvén velocity for strong guide fields (Shepherd and Cassak 2012).

The ultimate extent of the reconnection line must also be determined. In some models, reconnection is very localized (Russell and Elphic 1979). In others, both steady and sporadic reconnection occur along reconnection lines that extend over many hours in local time (Lockwood et al. 1990; Phan et al. 2000). A small amount of localized plasmaspheric mass-loading may redistribute the locations where reconnection occurs on the magnetopause, whereas large mass loading might cause system level reconfigurations (Zhang et al. 2016). Finally, reconnection may also occur simultaneously at numerous sites spread across broad regions of the dayside magnetopause (e.g., Alexeev et al. 1998), in which case different portions of the magnetopause might erode inward erratically in a disjointed manner. Distinguishing between these possibilities requires global images of the magnetopause.

Inferences concerning the location and thickness of plasma boundary layers just inside the magnetopause can also provide information concerning the location of magnetopause reconnection. Wave-driven diffusion, reconnection facilitated by nonlinear Kelvin-Helmholtz instabilities, and reconnection at remote locations can all produce such boundary layers (e.g., Nakamura et al. 2006; Hasegawa et al. 2009). By contrast to diffusion, which generates boundary layers whose thickness increases with distance downstream from the subsolar point, or Kelvin-Helmholtz instabilities, which generate boundary layers whose thickness depends on downstream distances and magnetopause velocity shear (i.e., solar wind velocity), reconnection produces boundary layers of mixed magnetosheath and magnetospheric plasma whose width increases with distance from the reconnection site (Sonnerup et al. 1981; Gosling et al. 1990). In soft X-ray images the presence of these boundary layers will be detected as a blurring of the plasma boundaries that would otherwise be present. The presence and absence of these boundary layers can therefore be used to determine when and where reconnection is occurring, thereby distinguishing between component, antiparallel, and other reconnection models, each of which predicts a distinctly different reconnection location as a function of solar wind conditions.

We know very little about what influence other solar wind parameters such as the Mach number, plasma beta, or solar wind dynamic pressure have upon the rate and mode of reconnection, but this could be readily discerned from both detailed case and statistical studies of magnetopause erosion employing global observations of the magnetopause location and motion for different combinations of solar wind parameters. For example, there are reasons to suppose that reconnection, magnetic flux erosion, and the cross polar cap potential drop all saturate for strong southward IMF orientations (Mühlbachler et al. 2005). Global simulations indicate a slowdown and stall in dayside magnetopause erosion, overdraped lobes that extend further sunward than the dayside magnetopause, less magnetotail flaring than would be expected based on an extrapolation of empirical models, and even a dimple on the subsolar magnetopause for large negative IMF B_z (Dmitriev and Suvorova 2000, 2012; Siscoe et al. 2004; Ober et al. 2002, 2006), all features that should be readily seen in global images. Thus, global images could be used to determine the precise combination of solar wind parameters (e.g., dynamic pressure and IMF B_z) when saturation sets in (Yang et al. 2003).

Elsen and Winglee (1997) predicted that the location of the subsolar magnetopause would exhibit a diminished response to IMF B_z as the solar wind pressure increases, and a diminished response to solar wind dynamic pressure as the southward component of IMF B_z increases. Using the limited *in situ* observations available for unusual combinations of solar wind parameters, both case (Shue et al. 1998, 2001) and statistical (Roelof and Sibeck 1993; Lin et al. 2010) studies suggest that erosion is indeed non-linear, i.e. that the radial distance to the dayside magnetopause does not vary linearly with IMF B_z, that the rate of erosion by IMF B_z diminishes for high solar wind dynamic pressures, and that the rate at which pressure changes compress the magnetosphere diminishes for strong southward IMF B_z. Global images could confirm, extend, and quantify these results. Since the magnetopause does not respond instantaneously to variations in the IMF orientation (or the solar wind dynamic pressure) it will almost certainly be necessary to include the time history of the IMF orientation in determinations of magnetopause location (Shue et al. 2000).

Images can also be used to identify the degree to which radial IMF orientations reduce pressure upon the dayside magnetosphere (Fairfield et al. 1990) and allow the dayside magnetopause to expand outward (Merka et al. 2003b; Suvorova et al. 2010; Dušík et al. 2010), perhaps in response to kinetic effects within the foreshock or to magnetohydrodynamic anisotropies (Samsonov et al. 2012, 2013, 2017). They can be used to detect the effects, if any, of dawn/dusk or spiral/orthospiral IMF orientations on the size and shape of the steady-state magnetosphere. Finally, although the waves (Kaufmann and Konradi 1969; Samsonov et al. 2015) generated by most solar wind discontinuities may sweep along the bow shock and magnetopause too rapidly to be tracked, soft X-ray images could be used to track the response of both boundaries to very oblique discontinuities, i.e., those which traverse the dayside magnetosphere very slowly because their normals lie nearly transverse to the Sun-Earth line (e.g., Takeuchi et al. 2002).

The magnetospheric magnetic field perturbations associated with the Region 2 and ring current systems enhance magnetic field strengths in the outer dayside magnetosphere and might therefore be expected to push the magnetopause outward (Schield 1969). Numerical simulations suggest that the subsolar magnetopause moves outward some 0.6 to 0.8 Earth radii when the ring current intensifies (Samsonov et al. 2016). However, theory (Tsyganenko and Sibeck 1994) and some empirical models (Petrinec and Russell 1993) indicate that the dayside magnetopause moves outward only slightly during intervals when the ring current is enhanced. Observations suggest that the duskside magnetopause may (Wrenn et al. 1981; McComas et al. 1993; Dmitriev et al. 2004, 2005, 2011; Dmitriev and Suvorova 2012) or may not (McComas et al. 1994) lie further from Earth than the dawnside magnetopause in response to an enhanced partial ring current.

2.1.2 The Earth's Cusps

Reconnection opens formerly closed magnetospheric magnetic field lines and allows solar wind mass, energy, and momentum to flow into the magnetosphere along bundles of open magnetic field lines that map from the magnetopause down to the high latitude dayside ionosphere (Heikkila and Winningham 1971). Plasma densities on these cusp magnetic field lines are slightly less than those in the magnetosheath, but far greater than those in the adjacent magnetosphere (Lavraud et al. 2004; Walsh et al. 2016a). Furthermore, the cusps extend deep into regions of the exosphere where neutral densities are very high. Consequently, the cusps must be bright soft X-ray emitters.

Because observations of the cusp are already available from both *in situ* (Escoubet et al. 1992; Pitout et al. 2006) and ground-based (Lockwood et al. 1989; Pinnock et al. 1993; Sandholt et al. 1998) observatories, one might ask why global images are needed. One answer is that it is difficult to extract complete comprehensive views of cusp behavior from the intermittent snapshots of *in situ* measurements along the paths followed by rapidly moving spacecraft. Another is that the spatially-limited optical views of the low-altitude cusp provided from a handful of stations in the northern and southern hemisphere tell us little about the cusp at mid- or high-altitudes. Global soft X-ray images will provide a broader view, one that connects our knowledge of magnetopause phenomena to the features seen on the ground. This section examines the wealth of information that can be learned about the solar wind-magnetosphere interaction from soft X-ray observations of the location, dimensions, motion, and structure of the Earth's cusps.

First consider the location of the cusps in local time. Both component and antiparallel reconnection predict reconnection along the equatorial magnetopause during intervals of strongly southward IMF orientation. Component reconnection may continue on the subsolar magnetopause during intervals of strong dawnward or duskward IMF orientation (Gonzalez and Mozer 1974), but antiparallel reconnection moves away from the subsolar point to off-equatorial locations (Crooker 1979). Although cusps produced by component reconnection may remain in place near local noon when the IMF has a dawnward or duskward IMF orientation, the antiparallel reconnection model predicts that duskward (dawnward) IMF orientations move the northern cusp duskward (dawnward) but the southern cusp dawnward (duskward). During periods of strong dawnward or duskward IMF orientation, reconnection may occur at both high and low latitudes, forming double cusps (Wing et al. 2001; Berchem et al. 2016). Soft X-ray observations of cusp locations will determine whether component or antiparallel reconnection prevails as a function of solar wind conditions.

Now consider the latitude of the cusps. During periods of southward IMF, enhanced reconnection rates on the dayside equatorial magnetopause cause the cusps to move $\sim 10°$ equatorward (Burch 1973; Carbary and Meng 1986; Wing et al. 2001). In the absence of simultaneous measurements in both hemispheres, we might suppose that the northern and southern hemisphere cusps move in unison to similar geomagnetic latitudes when the IMF turns southward. However, there is plenty of evidence indicating that their latitudes differ (Candidi and Meng 1988), for reasons that remain unclear. Global images with simultaneous solar wind coverage will afford an unprecedented opportunity to address this topic.

The response of the cusps to northward IMF orientations also remains to be fully established. Newell et al. (1989) reported observations indicating that reconnection moves to locations poleward of the cusp and appends magnetosheath magnetic field lines to the magnetosphere during periods of northward IMF orientation, causing the cusps to move poleward. By contrast, Palmroth et al. (2001) presented observations indicating equatorward cusp motion during intervals of strongly northward IMF and suggested that this might result from intensified reconnection on the equatorial magnetopause. Other work indicates that the latitudinal position of the high-altitude cusp does not move, but rather remains stationary for increasingly northward IMF orientations (Merka et al. 2002). Soft X-ray images will determine the latitudes of the cusps in both hemispheres as a function of time and discriminate between proposed models.

As cusp motion indicates the net rate at which magnetic flux is transferred from the dayside to the nightside magnetosphere (or vice-versa), determining the response time of the cusp to changing solar wind conditions and the velocity at which it moves is important to understand the state of the solar wind-magnetosphere interaction and time the development of storms and substorms. Yet the time scale for the cusp to respond to varying solar wind conditions remains unclear. Past observations indicate that the initial response begins almost immediately, but that a further 10 to 40 minutes are required to complete cusp relocations (Escoubet and Bosqued 1989; Němeček and Šafránková 2008). Yeoman et al. (2002) employed ground-based radar observations to track equatorward motion of the cusp during intervals of southward IMF orientation, but found no motion during intervals of northward IMF orientation. Pitout et al. (2006) reported several case studies in which snapshots from multipoint *in situ* observations indicated equatorward motion following southward IMF turnings, but poleward motion following northward IMF turnings. A wide field-of-view soft X-ray telescope will provide the sequences of images needed to identify cusp motion and time its velocity as a function of solar wind conditions. The observations could be used to determine whether or not steady-state conditions are ever achieved, how the magnetosphere responds to the onset of dayside and magnetotail reconnection, and how the magnetosphere responds to the cessation of dayside reconnection. Since the cusps lie at the boundary between open and closed magnetic field lines, observations of their latitude can immediately be used to quantify flux erosion from the dayside magnetosphere.

Just as in the case of the magnetopause, cusp motion can be steady, occur by leaps in response to individual southward IMF turnings (e.g., Lockwood et al. 1989), or occur by leaps in response to bursts of reconnection triggered by local magnetopause instabilities. The equatorward motion of the cusps may saturate for large southward IMF orientations (e.g., Siscoe et al. 2002; Ober et al. 2006). Little information is available concerning how the cusp moves in response to northward IMF turnings.

Now consider the response of the cusp to variations in the dipole tilt. Empirical models and both low- and high-altitude observations indicate that the cusps move equatorward in response to sunward diurnal and seasonal dipole tilts (Newell and Meng 1989; Zhou et al. 1999; Tsyganenko and Russell 1999). Both the width of the summer cusp and the densities within it exceed those of the winter cusp (Newell and Meng 1988; Pitout et al. 2006; Wiltberger et al. 2009). Simultaneous soft X-ray images of both cusps can be used to study these variations on a routine basis for the full range of solar wind and geomagnetic conditions, thereby quantifying how much plasma enters the magnetosphere in each hemisphere.

The width of the cusp yields important information concerning magnetospheric convection. The cusps span several Earth radii near the magnetopause (Walsh et al. 2012a) but narrow to dimensions of several hundred kilometers at their high-latitude, low-altitude, ionospheric footprints (Newell and Meng 1992). We adopt a kinetic interpretation to understand the internal structure of the cusps. The suprathermal magnetosheath particles entering the cusps precipitate into the high-latitude dayside ionosphere first, followed by the bulk of the distribution, and then the slower moving subthermal particles. Since the reconnected magnetic field lines within the cusps move in response to pressure gradient and magnetic field curvature forces, the precipitating particles exhibit distinctive spatial dispersion patterns (Rosenbauer et al. 1975; Reiff et al. 1977; Wing et al. 1996, 2001). The motion of magnetic field lines poleward from reconnection sites on the dayside equatorial magnetopause results in precipitating thermal and subthermal particle fluxes that initially increase abruptly and then subsequently decrease more gradually with latitude during periods of southward IMF orientation, as illustrated in Fig. 6. The width of the region over which they precipitate increases with increasing convection velocity. By contrast, during periods

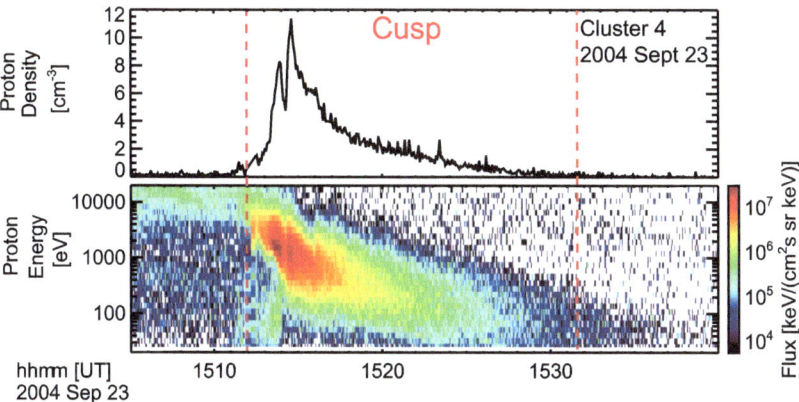

Fig. 6 Cluster 4 CIS instrument measurements of density structure in the cusp. The spacecraft cuts through the high altitude cusp from low to high latitudes, i.e., from GSM $(R, \lambda) = (4.62\ R_E, 54.5°)$ at 15:10 UT to $(4.84\ R_E, 64.5°)$ at 1530 UT. The plasma density peaks at the equatorward edge and gradually decreases with increasing latitude. Here R is the radial distance from Earth and λ is the latitude

of northward IMF orientation, newly reconnected magnetic field lines either stagnate or move equatorward. Precipitating particle fluxes should either increase with latitude or show little variation. During periods of dawnward or duskward IMF orientation, curvature and pressure gradient forces should pull the newly reconnected magnetic field lines azimuthally, resulting in dawn/dusk cusp particle dispersion patterns. All these features, and their time-dependencies, could readily be identified and quantified by a global imager.

The azimuthal extent of the cusp in the direction transverse to the convection velocity provides information concerning the extent of the reconnection line(s) on the dayside magnetopause. Broad cusps may map to a line 25 R_E long on the magnetopause for southward IMF orientations, but narrow cusps to a line only $\sim 5\ R_E$ long for northward IMF orientations (Fuselier et al. 2002). Azimuthal structure within the cusp can be interpreted as evidence for patchy reconnection on the dayside magnetopause. If reconnection occurs simultaneously along a single extended reconnection line, cusp properties will vary smoothly in azimuth. Whether or not it occurs simultaneously, patchy reconnection along multiple disconnected reconnection line segments will result in considerable azimuthal structure. Images of the cusp will provide information concerning the extent of reconnection on the dayside magnetopause.

Steady reconnection along a single reconnection line for either southward or northward IMF orientations should produce smooth variations in ion energy and density versus latitude. Stepped structures in the meridional direction (Newell and Meng 1991; Escoubet et al. 1992; Trattner et al. 2008) can therefore be interpreted as evidence either for time-varying reconnection (Smith and Lockwood 1990; Escoubet et al. 1992) or multiple reconnection sites at different latitudes (Kan 1988; Nishida 1989; Onsager et al. 1995; Trattner et al. 1999). Spatial and temporal variations can occur at the same time (Němeček et al. 2004). Soft X-ray images can be used to distinguish between these possibilities. Steady-state structures generated by multiple reconnection sites remain in place, whereas transient features produced by time-dependent reconnection convect antisunward. Images could also be used to determine the number and extent of such features, thereby addressing the locations of reconnection and the relative importance of steady and transient reconnection.

Finally, just as in the case of the magnetosphere as a whole, an increase in the solar wind dynamic pressure may diminish the dimensions of the cusp (Fung 1997). However, studies indicate that an increase in the solar wind dynamic pressure causes the dimensions of the cusp to increase (Zhou et al. 2000; Merka et al. 2002). Simulation results suggest that cusp dimensions initially increase with increasing solar wind pressure, but saturate near solar wind dynamic pressures of 3 nPa (Zhang et al. 2013). Perhaps the cusp widening results from greater magnetosheath magnetic field strengths and reconnection rates during intervals of enhanced solar wind dynamic pressure magnetopause (Newell and Meng 1994).

2.1.3 Transients at Earth's Magnetopause and in the Cusps

Transient structures/events with durations on the order of 30 s to several minutes are common in the vicinity of the Earth's magnetopause. They have been interpreted as the magnetospheric response to variations in the intrinsic solar wind dynamic pressure (Kaufmann and Konradi 1969), the magnetospheric response to transient dynamic pressure fluctuations generated within the foreshock (Fairfield et al. 1990), the Kelvin-Helmholtz instability operating at the magnetopause (Boller and Stolov 1973), and flux transfer events (FTEs) generated by bursts of magnetic reconnection between magnetosheath and magnetospheric magnetic field lines (Russell and Elphic 1978). If sufficiently numerous and extensive, the events might contribute significantly to (Lockwood et al. 1990) or even dominate (Lockwood et al. 1995) the solar wind-magnetosphere interaction. Consequently, quantifying the significance of each proposed transient solar wind-magnetosphere interaction mechanism as a function of solar wind conditions is a core objective of magnetospheric physics.

Although comprehensive single point and multipoint *in situ* measurements provide evidence for each of the proposed mechanisms, only instantaneous global measurements can definitively quantify their significance on the basis of their occurrence rates and dimensions. Fortunately, models for the various transient interaction mechanisms make very specific predictions concerning event occurrence patterns and signatures.

Solar wind tangential discontinuities are relatively common, arriving at Earth about once per hour (Burlaga and Ness 1969). Very few tangential discontinuities provide density variations greater than 35% (Solodyna et al. 1977). Although much rarer, interplanetary shocks often provide factor of two or larger density and dynamic pressure variations (e.g., Wang et al. 2010). Because they extend over many Earth radii transverse to the Sun-Earth line (Burlaga and Ness 1969), the pressure variations that accompany solar wind discontinuities launch widespread antisunward moving waves on the magnetopause. Transient enhancements in the solar wind dynamic pressure compress the magnetopause, while transient decreases allow it to expand outward. The same discontinuities launch fast mode waves that propagate throughout the magnetosphere. These fast mode waves may outrun the antisunward-moving solar wind discontinuities and initiate magnetopause motion ahead of the driving solar wind discontinuities. For example, the fast mode compressional waves launched by a transient increase in the solar wind dynamic pressure may cause the magnetopause to move outward in advance of the inward motion associated with the discontinuity itself (Kaufmann and Konradi 1969; Samsonov et al. 2015). The extent and amplitude of pressure-pulse induced waves could be determined by correlating global images of magnetopause motion with simultaneous *in situ* observations of solar wind dynamic pressure.

Kinetic effects in the foreshock generate more localized density and pressure variations (Thomas and Brecht 1988; Omidi and Sibeck 2007). Some (e.g. hot flow anomalies) lie centered on tangential discontinuities, others (e.g., foreshock cavities) are bounded by tangential discontinuities (Sibeck et al. 2001), some (e.g., compressional boundaries) bound the foreshock (Omidi et al. 2009), and yet others (e.g., spontaneous hot flow anomalies) lie within the foreshock but are not associated with discontinuities (Zhang et al. 2013). Corresponding

16

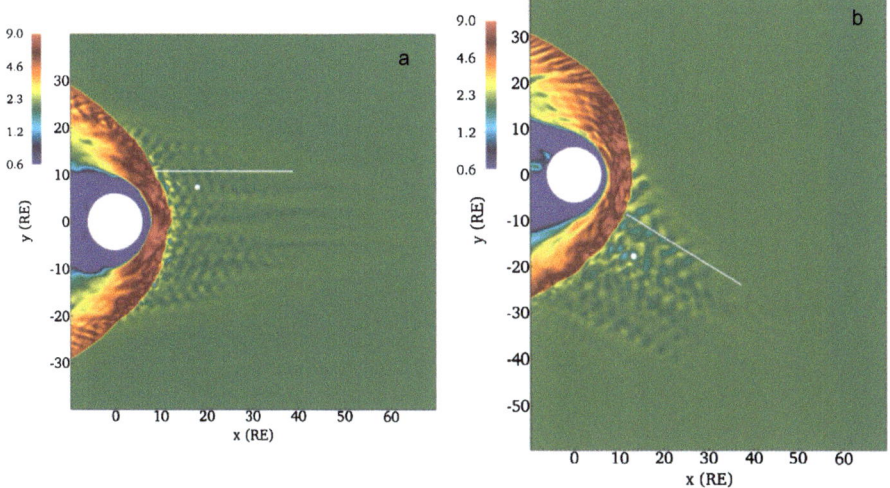

Fig. 7 Density structures resulting from kinetic processes within the foreshock (adapted from von Alfthan et al. 2014). The panels display density (cm^{-3}) from Vlasiator code hybrid-Vlasov simulations. Solar wind parameters are identical for the two panels, with the exception of the IMF orientation, which is radial in panel (a) but inclined 30° from radial in panel (b). The white line is parallel to the IMF orientation

ripples in the bow shock position result in magnetosheath plasma jets with enhanced densities capable of driving transient magnetopause motion and magnetospheric compressions (Hietala et al. 2012). The impact of these events on the magnetosphere should be greatest during intervals of radial or near-radial IMF orientation, as illustrated in the left panel of Fig. 7, when the IMF lies nearly along the Sun-Earth line and the foreshock lies upstream from the Earth's dayside magnetosphere (Fairfield et al. 1990). Because the foreshock lies upstream from the pre-noon bow shock and dayside magnetosphere for the typical spiral IMF orientation (see the right panel of Fig. 7), the magnetopause boundary waves and fast move waves transmitted into the magnetosphere by foreshock pressure pulses should generally be limited to the pre-noon magnetosphere (e.g., Howe and Binsack 1972; Rufenach et al. 1989; Russell et al. 1997). *In situ* observations indicate that the foreshock pressure pulses are more prominent during intervals of enhanced solar wind velocities (Sibeck et al. 2001; Facskó et al. 2008). Consequently, we expect the same to be true for the corresponding magnetopause motion. The significance of foreshock events can be determined by combining global images of magnetopause motion with *in situ* observations of solar wind variations, and in particular IMF orientations.

A Kelvin-Helmholtz instability occurs when flow shears at the magnetopause or inner edge of the low-latitude boundary layer overcome stabilizing curvature forces in draped magnetosheath and magnetospheric magnetic field lines and generate antisunward-propagating/convecting waves. The fastest growing wavelengths should be about 10 times greater than boundary layer thicknesses, with wave amplitudes increasing with increasing shears (Walker 1981) and downstream distance (Li et al. 2012). The instability is most likely to occur when strong flow shears lie perpendicular to both magnetosheath and magnetospheric magnetic field orientations, a condition most readily obtained on the equatorial flanks of the magnetosphere during intervals of strongly northward or southward IMF orientations (Southwood 1968). However, the instability can occur at other locations, including the high latitude magnetopause, when conditions are favorable (Hwang et al. 2012). It may be very common. Kelvin-Helmholtz waves occur about 40% of the time when the IMF points northward and about 10% of the time when it points southward. (Kavosi and Raeder 2015). Weaker magnetosheath magnetic field components parallel to the flow shear may make the instability more likely on the side of the magnetosphere behind the quasi-parallel bow shock (Nykyri 2013). We do not know if conditions sometimes favor hemispheric asymmetries in the occurrence of Kelvin-Helmholtz waves (Taylor et al. 2012) These predictions of the Kelvin-Helmholtz model can be tested by comparing global observations of magnetopause motion with *in situ* measurements of solar wind parameters.

Flux transfer events (FTEs) are bundles of intertwined magnetic field lines that, in contrast to the boundary waves generated by pressure pulses and the Kelvin-Helmholtz instability, simultaneously bulge outward into both the magnetosheath and magnetosphere. They contain a mixture of magnetosheath and magnetospheric plasmas, and consequently locally broaden and diminish the otherwise sharp density gradients that mark the magnetopause. Erkaev et al. (2003) attribute abrupt, pronounced, decreases and gradual increases of the density in the inner magnetosheath to bursts of reconnection. Because they result from reconnection, and reconnection is more likely when and where the shear between magnetosheath and magnetospheric magnetic field orientations is greater, FTEs on the dayside magnetopause are more common during intervals of southward IMF orientation (Berchem and Russell 1984). The origin of events on the flanks of the magnetosphere remains disputed. They may be generated by tilted reconnection lines that extend antisunward from the subsolar magnetopause (Kawano and Russell 1997), or be generated locally in regions of the high-latitude magnetopause where magnetosheath and magnetospheric magnetic field lines lie antiparallel (Sibeck et al. 2005). The fate of FTEs remains equally uncertain. They may slip over the polar magnetopause or be destroyed by interactions with magnetospheric magnetic field lines within the cusp regions (Omidi and Sibeck 2007). The occurrence of transients and FTEs may, or may not, be triggered by the arrival of solar wind discontinuities (Le et al. 1993; Lockwood and Wild 1993; Tkachenko et al. 2011). These and other questions could be readily answered with simultaneous global images of the magnetopause and *in situ* solar wind observations.

2.1.4 Earth's Magnetosheath

The magnetosheath envelops the magnetosphere, thereby providing its outer boundary conditions and the medium through which solar wind features are modified and transmitted to the magnetopause. Magnetosheath properties govern the occurrence patterns for reconnection and the Kelvin-Helmholtz instability at the magnetopause, which in turn control the flow of solar wind mass, energy, and momentum into the magnetosphere. In particular, low densities and low plasma beta favor the occurrence of magnetic reconnection (Phan et al. 2013), perhaps enabling steady reconnection to occur on high-latitude regions of the magnetopause where it would otherwise be precluded by high magnetosheath velocities (Fuselier et al. 2000; Avanov et al. 2001; Panov et al. 2008). By contrast, high densities and low Alfvén velocities favor the occurrence of the Kelvin-Helmholtz instability (e.g., Southwood 1968; Walsh et al. 2015).

Spreiter et al. (1966) reported the predictions of a gasdynamic model for an axially symmetric magnetosphere. Densities decrease slightly from the subsolar magnetopause to the bow shock along radial lines within $\sim 45°$ from the Sun-Earth line, but increase significantly from the magnetopause to the bow shock along radial lines at greater angles. MHD theory suggests that the presence of a magnetic field within the flowing plasma results in the formation of a plasma depletion layer (PDL) with very low magnetosheath densities but enhanced magnetic field strengths just outside the dayside magnetopause (Zwan and Wolf 1976). Numerical simulations indicate that stable depletion layers are present during intervals of steady northward IMF orientation. It can be difficult for individual spacecraft to detect the predicted smooth transitions and non-uniform increases in layer thickness with both latitude and longitude away from the subsolar point due to the back and forth motion of the layers in response to constantly varying solar wind plasma parameters (Wang et al. 2003). On the other hand, X-ray imagers should readily identify the appearance and disappearance of a PDL as a change in emission intensity and width of the magnetosheath to magnetosphere transition.

Modeling case studies suggest that the PDL can extend to cusp latitudes and 6 hours in local time away from noon (Wang et al. 2003). Predicted density depletion factors (for similar solar wind conditions) range from 1.2 (Lyon 1994) to 10 (Siscoe et al. 2002). Table 1 summarizes reported depletion layer dependencies on solar wind conditions. In some models, depletion factors and layer thicknesses diminish with increases in the IMF clock angle away from northward in the plane perpendicular to the Sun-Earth line (Siscoe et al. 2002), while in others they increase (Table 5 of Wang et al. 2004a; Pudovkin et al. 1995, 2001, 2002). Wang et al. (2004a) presented results from a parametric study indicating that depletion layer widths decrease with increasing solar wind magnetosonic Mach number, increase with increasingly northward IMF Bz strength, increase as the (clock angle) component of the IMF in the plane perpendicular to the Sun-Earth line rotates away from due northward, and remain almost constant as the dipole tilt increases. Wang et al. (2004a) also concluded that density depletion factors increase and then decrease as the solar wind magnetosonic Mach number increases, increase but then decrease as with increasingly northward IMF B_z strength, increase slightly as the IMF clock angle increases, and decrease as the dipole tilt increases. Simulation results presented by Maynard et al. (2004) indicate that depletion layers form just outside the dayside magnetopause even for southward IMF orientations when the IMF has a finite component along the Sun-Earth line and/or there is a strong dipole tilt, because reconnection moves to higher latitudes. Furthermore, they indicate depletion layers forming poleward of the cusps during intervals of very strongly southward IMF orientations.

Case and statistical studies of single-point *in situ* spacecraft observations provide support for some of these predictions. On the subsolar magnetopause, both layer thicknesses and depletion factors diminish when the IMF turns southward (Phan et al. 1994; Slivka et al. 2015) or radial (Anderson and Fuselier 1993). Farrugia et al. (1995) reported that the layer becomes more pronounced for low solar wind Mach numbers, while Anderson et al. (1997) reported a pronounced layer for high solar wind Mach numbers when the magnetosphere is compressed by high solar wind dynamic pressures, solar wind densities are large, and Alfvén velocities are low. Maynard et al. (2004) reported that the layer shifts to the region behind the quasi-perpendicular bow shock. Finally, a pronounced depletion layer has indeed been observed on the high latitude magnetopause during an interval of southward IMF orientation (Moretto et al. 2005). Contrary to model predictions, the depletion layer may become less prominent for small IMF cone angles (Anderson and Fuselier 1993). Soft X-ray images, like those proposed in this work, could be used to discriminate between these predictions and examine others yet to be tested.

Song et al. (1990) and Song and Russell (1992) reported observations of anticorrelated density enhancements and magnetic field strength depressions just upstream from the subsolar magnetopause and interpreted these observations in terms of standing slow mode waves. Southwood and Kivelson (1992, 1995) illustrated how a slow mode wave standing in the magnetosheath could result in a region with enhanced densities and depressed magnetic field strengths. Magnetic field lines within this region would have greater components parallel to the Sun-Earth line than those either further upstream in the magnetosheath proper

Table 1 Depletion layer predictions and observations

Property	Predicted dependence	Control parameter	Ref.	Observation
Thickness	Decreases with	Increasing magnetosonic Mach number	1, 2, 9	
	Increases with	Increasing positive IMF B_z	3	Present at subsolar magnetopause for IMF $B_z > 0$, absent for IMF $B_z < 0$ (Phan et al. 1994)
				Present for large negative IMF B_z on polar magnetopause (Moretto et al. 2005)
	Increases slightly with	Increasing IMF clock angle	3	
	Shows little effect along Sun-Earth line with	Increasing IMF cone angle	3	
	Is greater for	15° than 0° or 30° dipole tilt	3	
	Increases for	Increasing Sun-Earth-Observer Angle	4	
Density depletion	First decreases then increases as	Magnetosonic Mach number increases from 5.3 to 8.8	3	Larger for low Alfvénic Mach number (Farrugia et al. 1995) Larger for high Alfvénic Mach number (Anderson et al. 1997)
	First increases then decreases as	IMF B_z increases 2 to 21 nT	3	Much greater for $B_z > 0$ than $B_z < 0$ (Phan et al. 1994; Anderson et al. 1997)
	Increases non-monotonically as	IMF clock angle increases	3	Magnetic barrier (density depletion) increases with increasing clock angle (Pudovkin et al. 1995, 2001, 2002)
	Decreases as	IMF clock angle increases	5	
	Shows little effect along Sun-Earth line with	Increasing IMF cone angle	3	Becomes more prominent with increasing cone angle (Anderson and Fuselier 1993) Shifts to region behind quasi-perpendicular shock (Maynard et al. 2004)
	Decreases with	Increasing dipole tilt	3	
	Increases with	Increasing dipole tilt (or B_x)	6	
	Decreases with	Increasing Sun-Earth-Observer angle	4	
Enhanced densities in standing slow mode waves	Absent for all	IMF orientations	3, 7	Slow mode waves exist (Song et al. 1990; Song and Russell 1992)
	Present for non-zero	IMF cone angles	8	Should be interpreted as transmitted IMF discontinuities (Hubert and Samsonov 2004)

References: 1—Zwan and Wolf (1976); 2—Miura (1984); 3—Wang et al. (2004a); 4—Wang et al. (2003); 5—Siscoe et al. (2002); 6—Maynard et al. (2004); 7—Samsonov and Hubert (2004); 8—Lee et al. (1991); 9—Farrugia et al. (2000)

or downstream in the depletion layer. Lee et al. (1991) identified the anticorrelated features in two-dimensional incompressible MHD simulations whenever there was a magnetic field component parallel to the Sun-Earth line. However, Wang et al. (2004b,c) and Samsonov and Hubert (2004) were unable to find any such features in global MHD simulations for any IMF orientation. Hubert and Samsonov (2004) concluded that the anticorrelated density enhancements and magnetic field strength decreases were simply antisunward propagating solar wind features caught just before they encountered the magnetopause, which prompted a comment (Song et al. 2005) and reply (Hubert and Samsonov 2005). The issue remains unsettled, but could be addressed by imaging the structure of the inner magnetosheath.

Dawn/dusk asymmetries in magnetosheath densities may control the occurrence of reconnection and the entry of solar wind/magnetosheath plasma into the magnetosphere. This entry results in the formation of low-latitude boundary layers with magnetosheath-like plasma at densities lower than those in the magnetosheath. Observations indicating greater densities in the dawnside than duskside low-latitude boundary layer (LLBL, Hasegawa et al. 2003) and magnetotail plasma sheet (Wing et al. 2005) suggest greater pre- than post-noon magnetosheath densities and/or plasma entry. Walters (1964) argued that the presence of a Parker spiral IMF embedded in the flowing solar wind plasma would indeed result in greater dawnside than duskside magnetosheath densities, particularly during intervals of low solar wind Mach number. Global MHD models confirm this prediction for spiral IMF orientations, with asymmetries that increase for decreasing solar wind Mach number (Walsh et al. 2012b). Observationally, Paularena et al. (2001), Němeček et al. (2002), and Longmore et al. (2005) report greater densities in the dawnside magnetosheath than in the duskside magnetosheath. However, each of these studies concluded that the density asymmetry was unrelated to the IMF orientation. By contrast, Walsh et al. (2012b) reported asymmetries in the expected sense. A statistical survey reported by Dimmock and Nykyri (2013) found no evidence for any dawn/dusk density asymmetry, but Dimmock et al. (2016) went on to show that the expected asymmetries were in fact present, but only in the region immediately outside the magnetopause. Finally, note that greater densities and consequently enhanced plasma betas should inhibit reconnection. Rather than resulting from asymmetric magnetosheath densities, observations of enhanced densities in the dawnside boundary layer and plasma sheet may indicate the preferential operation of one or more diffusive entry processes.

2.1.5 Earth's Bow Shock

Simulations for the solar wind's interaction with the Earth's magnetosphere require accurate values for the polytropic index γ which represents the ratio of specific heats (C_p/C_v) and closes the set of magnetohydrodynamic equations. Determining the polytropic index is important because it controls phenomena as diverse as the degree of heating in magnetic reconnection (Hesse and Birn 1992) and magnetosheath flow deflections (Nishino et al. 2008). Theoretical values for γ range from 2 (for an adiabatic gas with two degrees of freedom perpendicular to the magnetic field), through 5/3 (for an adiabatic gas with three degrees of freedom), 1.5, and 1.33 (when there is a heat flux escaping from the magnetosheath into the solar wind, Nishino et al. 2008), to 1 (for an isothermal gas). Observationally-inferred values for γ are almost as diverse, ranging from 1.67 (Russell et al. 1983), through 1.76 (Farris et al. 1991) and 1.85 (Tatrallyay et al. 1984), to 2 (Zhuang and Russell 1981).

Density jumps at the bow shock provide crucial information concerning γ (Farris et al. 1991). Following Spreiter et al. (1966), the jumps are a function of both γ and the upstream solar wind magnetosonic Mach number (MMS), i.e., $\rho/\rho_{sw} = (\gamma + 1)M_{MS}^2/[(\gamma-1)M_{MS}^2 + 2]$, where ρ is the density in the subsolar magnetosheath. For typical values of

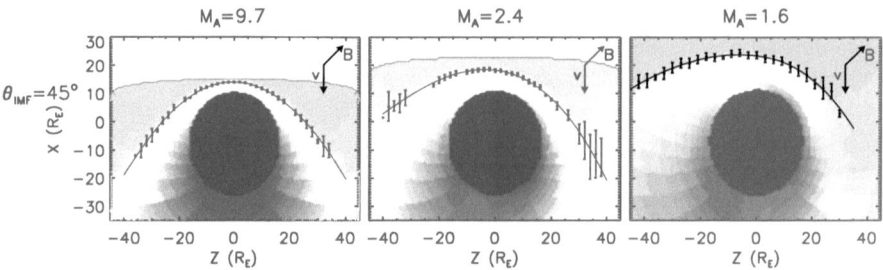

Fig. 8 Bow shock locations and shapes predicted by MHD models for three solar wind Alfvénic Mach numbers (adapted from Chapman et al. 2004). The IMF lies 45° from the Earth-Sun line

$M_{MS} \gg 1$, ρ/ρ_{sw} approaches M_{MS}^2 for $\gamma = 1$, 4 for $\gamma = \frac{5}{3}$, and 3 for $\gamma = 2$. Alternatively, the locations of the bow shock and magnetopause themselves can also be used to determine γ. As noted by Farris et al. (1991), $\gamma = [(1.1 + \Delta/D)M_{MS}^2 - 2.2]/[(1.1 - \Delta/D)M_{MS}^2]$, where D is the standoff distance of the magnetopause from the center of the Earth and Δ is the distance between the bow shock and the magnetopause. There are alternative formulations within the gasdynamic framework, including those that take into account the possibility that the density jump approaches unity and the bow shock recedes to infinity as the Mach number approaches unity, or that the solar wind feels the effects of the magnetospheric shape rather than the distance between the bow shock and the magnetosheath (Farris and Russell 1994). There are also magnetohydrodynamic approaches (Cairns and Grabbe 1994; Grabbe and Cairns 1995). When the Mach number approaches unity, Alfvén wings may form, greatly modifying the size and shape of the magnetopause (Ridley 2007; Chané et al. 2012).

The various models make strikingly different predictions for the location of the subsolar bow shock as a function of IMF orientation and solar wind Mach number. Cairns and Lyon (1996) predicted that the standoff distance increases as the solar wind Mach number decreases for IMF orientations transverse to the Sun-Earth line, but decreases for IMF orientations parallel to the Sun-Earth line. Models presented by Cairns and Grabbe (1994) and Cairns and Lyon (1996) predict standoff distances for low solar wind Mach numbers far greater than those predicted by Verigin et al. (2001) or Farris and Russell (1994). As illustrated in Fig. 8, the quasi-perpendicular bow shock lies further upstream than the quasi-parallel bow shock, with the discrepancy increasing as the solar wind Mach number decreases (Chapman and Cairns 2004). The latter authors predict a dimple on the subsolar bow shock for very low solar wind Mach numbers and radial IMF orientations.

It has proven difficult to verify these predictions with studies employing *in situ* observations. Despite multipoint observations, Fairfield et al. (2001) was unable to discriminate between the models for an unusually distant bow shock for low solar wind Mach numbers. Consistent with expectations, Slavin et al. (1996), Merka et al. (2003b), and Jelínek et al. (2010) reported subsolar bow shock locations closer to Earth and therefore very thin subsolar magnetosheaths during intervals of radial IMF orientation. Verigin et al. (2001) reported results from a small statistical study indicating that the standoff distance to the bow shock increases with increasing Alfvénic Mach number for field-aligned solar wind flows, but decreases for non-field-aligned flows. However, Jeřáb et al. (2005) could find no dependence of the bow shock location upon the IMF orientation whatsoever. Jeřáb et al. (2005) attributed the absence of any inward bow shock motion associated with southward IMF turnings and inward magnetopause erosion to a compensatory increase in the magnetosheath thickness associated with the blunter obstacle posed by an eroded dayside magnetopause. Nevertheless, Jeřáb et al. (2005) did find that the distance to the bow shock increases linearly as a function of the IMF strength.

One might expect abrupt variations in solar wind parameters to drive corresponding inward and outward motions of the bow shock and magnetopause (e.g., Fairfield et al. 2001). Consistent with this hypothesis, Anderson et al. (1968) found a good correspondence between the periods and amplitudes of bow shock and magnetopause motion on individual spacecraft passes. However, the similarity of the periods and amplitudes does not necessarily mean the two boundaries move inward and outward in phase. Korotova et al. (2012) recently reported that the same change in the IMF orientation drove transient outward bow shock motion, but transient inward magnetopause motion. And Jelínek et al. (2006) reported that the motion of the bow shock does not correspond to that of the magnetopause.

Summarizing results to date, Merka et al. (2003a, 2005) noted that existing models for the bow shock underestimate the distance to the bow shock under strong IMF conditions, fail to reflect the effects of variations in the IMF and solar wind velocity vectors, and do not correctly describe the bow shock location during intervals of low solar wind Mach number. Even large statistical studies based on *in situ* observations fail to resolve expected dawn/dusk differences and Mach cone asymmetries. Global images of the bow shock and magnetopause should be able to resolve these and other issues by identifying the locations of the bow shock and magnetopause, determining the density jump at the bow shock, discriminating between models, and providing the information needed to determine γ.

2.1.6 Earth's Foreshock

The magnitude of the jump in magnetic field strengths (or densities) at the bow shock determines its ability to accelerate particles. Shock-drift acceleration at the quasi-perpendicular bow shock produces beams of ions and electrons on magnetic field lines that lie perpendicular to the bow shock normal (Decker 1983). The maximum energy gained by the reflected particles is given by $T_f/T_i = 2r[1 + (1 - r^{-1})^{\frac{1}{2}}] - 1$, where T_i and T_f are the initial and final particle energies and r is the ratio of the magnetosheath to IMF strengths. Solar wind ions with ~ 1 keV energies might be accelerated to ~ 14 keV for $r = 4$. By contrast Fermi acceleration of an incident monoenergetic particle distribution at the quasi-parallel bow shock can produce diffuse ion populations with far greater energies, near-isotropic pitch angle distributions, and power law spectra whose spectral indices depend upon the ratio of magnetosheath to interplanetary magnetic field strengths. In the non-relativistic case, the steady-state spectral index for the distribution function is given by $3r/(r-1)$ (Blandford and Ostriker 1978).

Soft X-ray images can provide the information needed to determine the extent and nature of particle acceleration at the bow shock. First, the images can be used to identify the transition between the quasi-parallel and quasi-perpendicular bow shocks, which is expected to occur where the angle between the IMF and the normal to the bow shock passes through $45°$. A sharp density discontinuity indicates the quasi-perpendicular bow shock, whereas a broader and far more turbulent transition should mark the quasi-parallel bow shock. Secondly, the images can be used to determine the strength of the density, and consequently the magnetic field strength, jump at the bow shock.

Kinetic effects generate a wealth of mesoscale density structures upstream from Earth's bow shock, including hot flow anomalies (Thomsen et al. 1986), foreshock cavities (Sibeck et al. 2001), density holes (Parks et al. 2006), and bubbles (Turner et al. 2013). By enhancing and/or diminishing upstream densities, deflecting solar wind flows, and perturbing corresponding magnetosheath parameters, these structures generate prominent transient events in the outer dayside magnetosphere and dayside auroral ionosphere. However, with

Fig. 9 Schematic representation of the global morphology of the solar wind interaction with a cometary atmosphere, showing the various discontinuities in the flow pattern (adapted from Mendis 1988)

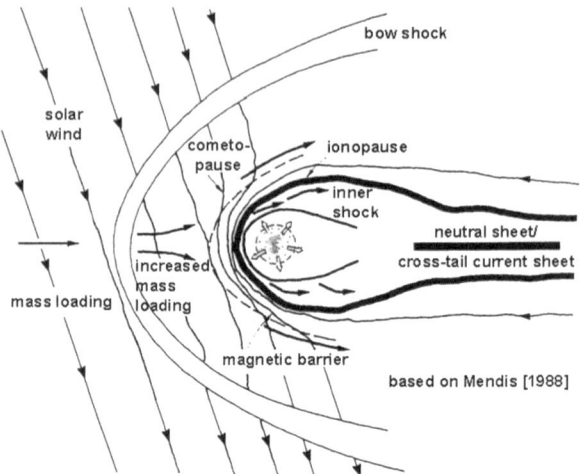

one exception, the limited dimensions and ephemeral nature of most of these features and their magnetospheric responses probably preclude soft X-ray imaging. The exception is the foreshock compressional boundary, a region of enhanced density piled up on the edges of the quasi-parallel foreshock (Omidi et al. 2009, 2013). Numerical simulations indicate that these structures can be quasi-steady-state features for a wide variety of IMF orientations. Some observational studies support this point of view, while others interpret the density enhancements as foreshock-generated structures moving antisunward with the solar wind flow (Sibeck et al. 2008; Billingham et al. 2008, 2011). Since the density enhancements and depletions associated with the structures extend nearly normal to the bow shock, it should be relatively easy to employ global images to distinguish between these two models.

2.2 Comets

Many visible light observations and a handful of *in situ* measurements provide tantalizing views of the complex plasma phenomena that occur when the solar wind encounters comets. As shown in Fig. 9, *in situ* observations indicate that these structures include a bow shock, a "cometopause", and an ionopause (Mendis 1988; Flammer 1991; Mendis and Horányi 2014). The bow shock forms in response to mass loading. As they approach the Sun, comets sublimate large clouds of neutral gas. The solar wind flow picks up ionized atoms and molecules within this cloud. If sufficiently numerous, the pick-up ions slow the flow down to the point where a bow shock forms. Deeper inside the bow shock, a "collisionopause" or "cometopause" forms at the transition from the heated and decelerated shocked collisionless mass-loaded flow to flow cooled and even more significantly decelerated by collisions and charge exchange with expanding cometary neutral molecules, in addition to pick-up ions generated by photoionization. Still closer to the nucleus lies the ionopause, the locus of points where the solar wind plasma makes its closest approach to the comet.

In situ observations confirm the presence of weak shocks on the nightside flanks of cometary tails (Coates 1995). Dayside shock strengths, and corresponding density enhancements, should be far greater. Simulations demonstrate that the IMF orientation controls the nature and thickness of the bow shock (Omidi and Winske 1987) which is thin for quasi-perpendicular configurations, broader for an intermediate shock, and narrow again for quasi-parallel shocks. Theory indicates that the dimensions of the bow shock increase as the

Fig. 10 Shading shows the soft X-ray intensities of Comet Hyakutake observed by the ROSAT Wide Field Camera (WFC), while contours show the intensity of the best adapted hydrodynamic model (Wegmann et al. 2004). The nucleus lies at $(X, R) = (0, 0)$, where X points towards the Sun. Lengths are in units of 10^5 km

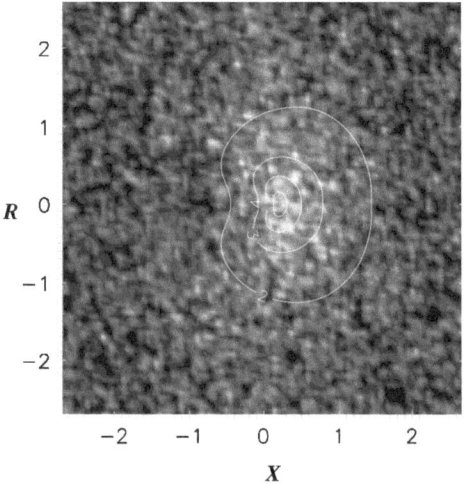

solar wind Mach number diminishes and/or gas production rate increases. Recent models show that the dimensions of the bowshock increase with greater photoionization, charge-exchange, and electron impact ionization (Simon Wedlund et al. 2017). Consequently, the dimensions of the magnetosheath region of shocked solar wind plasma behind the dayside bow shock should also increase as comets move sunward and sublimation increases (Flammer 1991). Furthermore, since the rate of ion pick up via microscopic wave-particle interactions should be lower than that via macroscopic motional $\mathbf{E} \times \mathbf{B}$ electric fields, a quasi-parallel bow shock should lie nearer the comet than a quasi-perpendicular bow shock (Omidi and Winske 1986).

The cometopause separates fast moving shocked solar wind flow from a region dominated by compressed IMFs and cometary ions. The width of the cometopause, where solar wind densities diminish, may be abrupt ($\sim 10^4$ km) (Gringauz et al. 1986) perhaps in response to a charge exchange avalanche (Gombosi 1987), or more diffuse (Balsiger et al. 1986; Amata et al. 1986), with the width depending upon the IMF orientation (Galeev et al. 1988). Within the cometopause, the solar wind proton flow decelerates rapidly and cools in response to charge exchange with cometary neutrals. Correspondingly, the densities of both major (H^+, He^{++}) and minor (e.g., O^{6+}, C^{5+}) solar wind species should increase. Finally, no solar wind ions reach locations closer to the comet than the ionopause, although comets with sufficiently low outgassing rates may lack an ionopause. As in the case of the bow shock, theory predicts that the cometopause and ionopause structures will move outward as comets approach the Sun and sublimation increases (Flammer 1991).

Optical emission in cometary tails is dominated by molecular band emission due to CO^+ and H_2O^+ while the in the cometary head, it is dominated by molecular band emission from C_2 and the reflection of sunlight by dust. In contrast to optical observations, soft X-ray observations (e.g., Lisse et al. 1996; Dennerl et al. 1997; Gao and Kwong 2002) are dominated by interactions with water and its dissociation products OH, O, and H (Bodewits et al. 2007) and can be used to determine the characteristics of the dayside plasma and neutral environments of comets on a routine basis as a function of solar wind conditions and distance from the Sun. As illustrated in Fig. 10, theory predicts and observations confirm that the attenuation of solar wind ion densities via charge transfer collisions with increasing depth into the extended cometary atmosphere or coma results in integrated LOS soft X-ray emissions that peak in a bowl-shaped region within the magnetosheath on the sunward side of cometary nuclei (Wegmann et al. 2004). (However, there are cases such as 2P/Encke (Lisse et al. 2005), where the coma is of sufficiently low density that it is collisionally thin to charge-exchange, in which case the morphology is roughly spherical.) Soft X-ray emissions should be far greater in the magnetosheath than in the solar wind thanks to greatly enhanced plasma densities and thermal velocities in the magnetosheath, as well as greater neutral densities. Beyond the bow shock, emissions should fall off as an inverse function of radial distance from the nucleus. Both the intensity and dimensions of the emitting region depend on the rate of neutral gas production, and should therefore increase as comets approach the Sun. Individual line intensities also depend upon the flux of high charge state ions, which varies with the state of the solar wind and the concomitant ion abundances (Bodewits et al. 2007). Finally, the Kelvin-Helmholtz instability may locally permit solar wind plasma to penetrate deep into cometary ionospheres as predicted by Ershkovich and Mendis (1983) and as seemingly observed at 67P/Churyumov-Gerasimenko by Goetz et al. (2016a,b).

2.3 Mars and Venus

Identifying and quantifying the processes that cause atmospheric loss is a major objective of planetary studies. They were the principle objectives of NASA's recent MAVEN mission to Mars (Jakosky et al. 2015b), and important objectives of ESA's Mars (Chicarro et al. 2004) and Venus (Titov et al. 2006) Express missions.

A number of processes govern the loss of the Martian and Venusian atmospheres (e.g., Nagy et al. 2004; Lammer et al. 2006; Dubinin et al. 1996; Lundin 2011). Some invoke bombardment and hydrodynamic outflow. Others involve solar wind interactions with the planetary atmosphere and/or ionosphere, such as the removal of pick up ions generated by photoionization or charge exchange, or the formation of detached blobs of ionospheric plasma generated by either the Kelvin-Helmholtz instability at the ionopause or magnetic reconnection with ionospheric or remnant crustal magnetic fields. Even in the absence of solar wind stripping, ambipolar electric fields may cause a planet to lose heavy ions (Collinson et al. 2015, 2016).

Although there is evidence for enhanced escape of ionospheric ions during space weather storms (Luhmann et al. 2007; Edberg et al. 2011; Collinson et al. 2015; Jakosky et al. 2015a), assessing the significance of these and other mechanisms with isolated *in situ* measurements can be difficult. Models can help (Lillis et al. 2015), but models for Mars predict escape fluxes that differ by more than an order of magnitude (Brecht and Ledvina 2006; Brain et al. 2010b). Observations indicate similar variations over the course of a solar cycle (Lundin et al. 2013), which must be due at least in part to the large variations in exospheric densities that occur over the solar cycle (e.g., Forbes et al. 2008). The significance of each mechanism that invokes solar wind-planetary interactions can be quantified using global images of the corresponding diagnostic plasma density structures.

As in the case of comets, the interaction of the supersonic solar wind with Mars generates several plasma structures where the densities of ions with solar wind origin change abruptly (Brain 2006; Bößwetter et al. 2007). Some of these boundaries are illustrated in Fig. 11, while densities from a numerical simulation are shown in Fig. 1. The outer edge of the magnetosheath is bounded by a bow shock where densities increase abruptly from solar wind to magnetosheath values. The density of ions diminishes gradually from the magnetosheath to the ionospheric side of the magnetic pile-up region (MPR) on the inner edge of the magnetosheath, in a manner akin to that in the depletion layer outside Earth's magnetopause. Electron observations indicate that this boundary is either lumpy or permeable in regions of radial crustal magnetic fields (Brain et al. 2005). Finally, the foreshock lies upstream from

Fig. 11 Cartoon of the global Martian solar wind interaction (Brain 2006). Orange shading indicates the density of planetary neutrals. Blue indicates the relative density of solar wind ions in different plasma regions (labeled in black), separated by different plasma boundaries (labeled in magenta). Here MPR stands for the magnetic pileup region, MPB for magnetic pile up boundary, and PEB for the photoelectron boundary

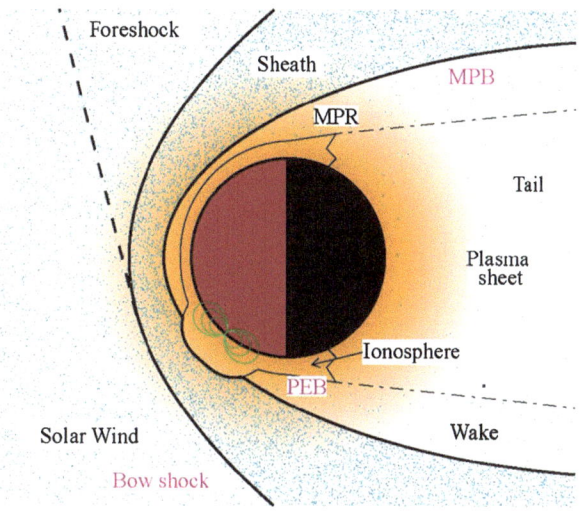

Table 2 Drivers affecting the variability of plasma boundary locations at Mars (Bertucci et al. 2005; Zhang et al. 1991a,b)

	Bow shock	Magnetic pile-up boundary	Photoelectron boundary
Solar wind pressure	?	Yes	Yes
IMF direction	Yes	Yes	?
EUV	?	?	Yes
Martian season	?	Yes/?	?
Crustal fields	No/?	Yes	Yes

the bow shock on IMF lines connected to that boundary. By analogy to Earth, we expect regions of enhanced solar wind densities to bound a foreshock exhibiting depressed densities.

Table 2 (Brain 2006) summarizes the reported effects of the solar wind dynamic pressure, IMF orientation, solar extreme ultraviolet (EUV) radiation, season, and crustal magnetic fields on the distances to the bow shock, magnetic pile-up boundary, and photoelectron boundary at Mars. Increases in the solar wind dynamic pressure may (Brain et al. 2005; Crider et al. 2005; Morgan et al. 2014) or may not (Trotignon et al. 1996) move the magnetic pile-up and photoelectron boundaries towards the planet. The terminator bow shock lies further from the planet in the directions perpendicular to the IMF orientation (Zhang et al. 1991a). The Martian bow shock flares and moves further from the planet as the solar wind Mach number decreases (Edberg et al. 2010).

Crustal magnetic fields affect plasma and magnetic field structures in the vicinity of Mars. They raise the distances to the magnetic pile-up and photoelectron boundaries, thereby locally precluding direct solar wind interactions with the ionosphere, but they do not appear to increase the distance to the bow shock (Brain 2006, and references therein). The IMF orientation may also control the altitude of the magnetic pile-up boundary. Mars Surveyor observations suggest that the altitude of the pile-up boundary rises for eastward IMF orientations but falls for southwest IMF orientations when the subsolar latitude lies in the northern hemisphere (Brain et al. 2005). Without global observations, it is difficult to determine whether this variation results from some as yet unspecific global cause, a local Hall current

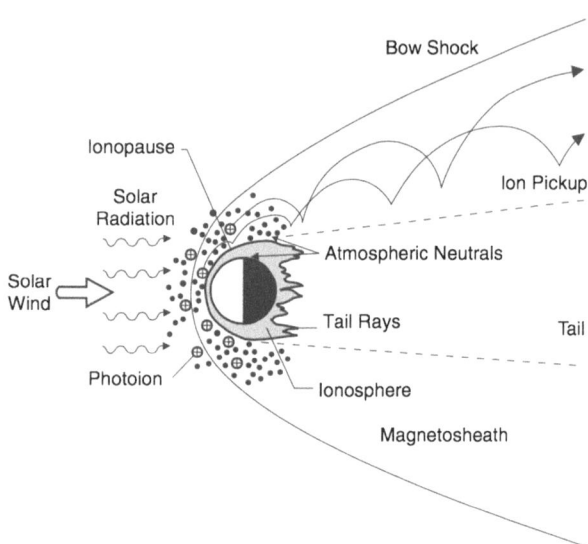

Fig. 12 Schematic illustration of the solar wind interaction with Venus. Solar EUV radiation ionizes the neutral upper atmosphere of Venus. The electron ion thermal pressures suffice to stand off the supersonic solar wind and form a shock. Neutrals formed in the flowing solar wind are carried away by the wind (Russell et al. 2006)

effect, or local mass loading. The altitude of the magnetic pile-up boundary increases in southern summer, when the stronger crustal magnetic fields in the southern hemisphere approach the subsolar point. Crustal anomalies may also determine magnetosheath densities. Ma et al. (2002) reported simulation results indicating that crustal magnetic features do not cause major distortions of the bow shock, but do have effects on the magnetosheath and the altitude of the ionopause. Simulation results reported by Harnett and Winglee (2003, 2005) predict that mini-magnetospheres extend beyond and replace the magnetic pile up boundary in the presence of crustal anomalies. In the absence of reconnection, IMF draping over strong southern magnetic anomalies at Mars should enhance flank magnetosheath densities by more than a factor of 2 outside dawnward or duskward facing anomalies. In the presence of reconnection, enhancements are far smaller and densities fall within void regions that lie just downstream from the anomalies. According to Vignes et al. (2000) and Bertucci et al. (2005), the bow shock and magnetic pile-up boundaries move inward and outward together, but neither boundary exhibits much response to variations in solar EUV. However, Brain et al. (2005) found evidence for the pile-up boundary moving towards the planet during periods of enhanced solar EUV.

Russell et al. (2006) presented the schematic illustration of the solar wind's interaction with Venus shown in Fig. 12 and enumerated the plasma structures seen by Pioneer Venus Orbiter (PVO, Colin 1980) at Venus. As in the case of Mars, the interaction results in the formation of a foreshock upstream from the quasi-parallel bow shock, a magnetosheath in which the plasma flow and magnetic field pick up planetary ions, a magnetic barrier with a mixture of solar wind and ionospheric plasmas at the inner edge of the magnetosheath, a (generally) field-free ionosphere, and a mass-loaded magnetotail. Because gradients in the densities of ions with solar wind origins mark each of these boundaries (e.g., Terada et al. 2009), they are imageable in soft X-rays. Table 3 summarizes reported effects of the solar wind pressure/Mach number, IMF direction, and solar EUV on plasma structures at Venus. The ionopause at the inner edge of the magnetic pile up boundary rises from altitudes of 300 to 1000 km as the solar wind dynamic pressure diminishes from 4 to 0.5 nPa (Brace et al. 1982). Distances to the bow shock and the width of the magnetosheath depend primarily on the IMF orientation, the solar wind Mach number, and exospheric neutral densities, rather

Table 3 Drivers affecting the variability of plasma boundary locations at Venus (Phillips et al. 1985; Knudsen et al. 1987; Russell et al. 1988; Brace et al. 1990)

	Bow shock	Magnetic pile-up boundary	Ionopause/photoelectron boundary
Solar wind pressure/	?	?	Yes
Mach number	?	?	Yes
Mach number	Yes	?	?
IMF direction	Yes	?	?
EUV	Maybe	?	Yes

than solar wind dynamic pressure (Martinecz et al. 2008; Russell et al. 1988; Zhang et al. 2004). Note, however, that Martinecz et al. (2009) found no relation between the location of the terminator bow shock and solar EUV. In the absence of a significant crustal magnetic field, there are no seasonal or diurnal effects.

Slavin et al. (1980) and Tatrallyay et al. (1983) showed that the bow shock at Venus flares more than might be expected based on gasdynamic models, suggesting that mass loading plays an important role. Consistent with this hypothesis, Alexander and Russell (1985) showed that the terminator bow shock moves outward during solar maximum when exospheric neutral densities should be enhanced. Alexander et al. (1986) demonstrated that the bow shock moves away from the planet as the IMF rotates from orientations parallel to the solar wind flow to orientations perpendicular to that flow, i.e., from orientations that do not favor ion pick up to orientations that do. The effect is far greater during solar maximum than solar minimum. As in the case of Mars, the bow shock moves outward for low Mach numbers (Russell et al. 1988). The Venusian bow shock is not circular within the terminator plane, but rather lies further from the planet in the direction perpendicular to the IMF orientation in the plane transverse to the solar wind flow direction, particularly in the direction with the outward pointing electric field where pick up ion effects are expected. Also consistent with the pick up ion effect, this asymmetry in bow shock locations becomes more pronounced during solar maximum. Finally, Zhang et al. (1990) used observed and estimated bow shock locations to infer that the effective radius of the Venusian obstacle to the solar wind lies below the distance to the subsolar ionopause during solar minimum, i.e., that there is a more direct interaction of solar wind plasma with this planet's ionosphere and exosphere during solar minimum than solar maximum.

The locations of the bow shock at Venus can be used to determine the best value for the polytropic index in the solar wind. Tatrallyay et al. (1984) discussed the strength of the Venusian bow shock by determining magnetic field strength compressions across the bow shock as a function of solar wind Mach number, concluding that polytropic index $\gamma = 1.85$ works best at Venus. The strength of the compressions, and the index γ, increase with magnetosonic Mach number and cone angle. The distance to the terminator shock diminishes with magnetosonic Mach number.

Hybrid code simulations predict the principle features of the solar wind's interaction with unmagnetized planets including the locations of the bow shock and an ion composition boundary between plasma of solar and planetary origin (Bößwetter et al. 2007), but have also made some interesting predictions for the solar wind interactions with Venus and Mars. Whereas simulations predict that the bow shock lies further from the planet in the hemisphere where the convection electric field points inwards towards the planet (e.g., Modolo et al. 2006; Brecht and Ledvina 2006), observations indicate that the bow shock lies further from the planet in the hemisphere containing accelerated pick-up ions, i.e., the hemisphere where the electric field points outward away from the planet (e.g., Dubinin et al. 1996, 1998; Vignes et al. 2002). Shimazu (2001a) reported that the sense of bow shock asymmetries could be reconciled with observations by including the effects of charge exchange in the models. In this case, heavy ions replace the flow of solar wind ions in the magnetosheath. Simulations also predict the occurrence of multiple shock waves (Moore et al. 1991; Shimazu 2001b; Modolo et al. 2006), a feature that would be difficult to identify using single point observations from *in situ* spacecraft. Because the model results reported by Moore et al. (1991) were relatively insensitive to mass-loading, these authors proposed that solar cycle variations in shock locations result from changes in the dimensions of the magnetic pile up boundary and ionopause rather than changes in the rate of ion pick up. Shimazu (2001b) predicted that the presence of the interplanetary magnetic field constrains the planetary plasma boundary (the ionopause) to an elliptical cross-section. Finally, Martinecz et al. (2009) predicted that pronounced density enhancements extend upstream from the quasi-parallel bow shock at the dawn terminator during intervals of spiral interplanetary magnetic fields. The density jump at the dawn terminator bow shock is much less than that on the dusk side. Brain et al. (2010b) compare the differing predictions of MHD and hybrid models for the solar wind interaction with Mars.

Transient plasma and magnetic field structures are common at both Mars and Venus. Some of these structures might be produced by external solar wind/foreshock drivers, while others result from instabilities at internal plasma boundaries. Within the former category, Collinson et al. (2014) suggested that the significant transient density structures generated by kinetic processes within the Venusian foreshock might drive large amplitude waves on the ionopause and magnetic pile-up boundary of that planet. The same might also be true at Mars, where hot diamagnetic cavities and flow anomalies are also present within the foreshock (Øieroset et al. 2001; Collinson et al. 2015).

However, instabilities may generate other structures. Observations indicate not only the common occurrence of magnetic flux ropes embedded in the ionosphere (Russell and Elphic 1979) but also the frequent occurrence of wavelike structures at the ionopause and clouds of ionospheric plasma above it (Brace et al. 1980, 1982; Acuna et al. 1998). Slowly moving flux ropes are common in the Venusian ionosphere during periods of low solar wind dynamic pressure at solar maximum (Russell et al. 2006), and they are present, albeit much rarer, at Mars (Vignes et al. 2004). According to some estimates these transient events may play a major role in removing planetary plasma (Brace et al. 1982; Russell et al. 1982b; Terada et al. 2002).

The clouds might be caused by a sling shot effect pulling draped magnetic field lines over the planetary poles (Russell et al. 1982b), reconnection and the formation of flux ropes in the magnetosheath (Dreher et al. 1995), or a tearing off of blobs during the final stage of a non-linear Kelvin-Helmholtz instability at the ionopause (Wolff et al. 1980; Gunell et al. 2008). Seeking to confirm the slingshot effect model, Ong et al. (1991) found that clouds are more common on Pioneer Venus Orbiter periapsis passes during which the orientation of the upstream magnetic field changes abruptly, indicating a need for an additional mechanism.

One such mechanism is magnetic reconnection, which should produce magnetic flux ropes in regions of sheared draped magnetosheath magnetic fields outside the ionopause (Dreher et al. 1995). Sheared fields are indeed a natural consequence of rotations in the upstream IMF orientation. However, at Mars there is another way that reconnection can generate flux ropes. Ma et al. (2002) reported simulation results indicating that reconnection of IMF and crustal magnetic field generates mini magnetocylinders of closed magnetic field lines within the Martian magnetosheath, while Harnett (2009) reported simulation results indicating that these cylinders are rapidly dissipating flux ropes with sizes that increase

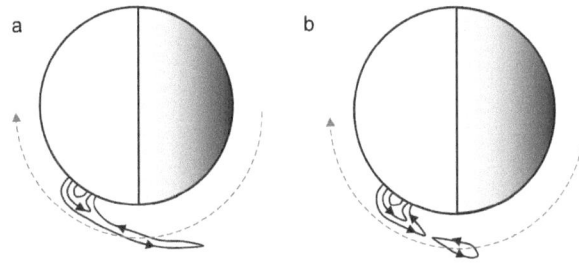

Fig. 13 Detaching a magnetic flux rope from Mars (Brain et al. 2010a). Panel (a) shows crustal magnetic field lines that are still attached to the planet, but have been stretched tailward by the solar wind. Panel (b) shows a detached loop of crustal magnetic field carrying plasma away from Mars. The dashed line shows the sunward motion of a spacecraft

slightly and are on the order of half a planetary radius in the Martian magnetosheath. Brain et al. (2010a) discuss observations indicating that interactions with the solar wind stretch and pinch off loops of crustal magnetic field, resulting in antisunward moving flux ropes filled with ionospheric plasma as shown in Fig. 13. Ropes with greater than 100 nT magnetic field strengths were seen in 1% of Mars Global Surveyor orbits and their estimated diameters were on the order of 2250 km, large compared to the radius of Mars. Consequently, Albee et al. (2001) argued that the ropes might account for up to 5–10% of ion loss at Mars.

Reconnection is not the only mechanism for generating flux ropes. Although the subsolar ionopause is generally thought to be stable to both the Kelvin-Helmholtz and flute instabilities (Elphic and Ershkovich 1984), large corrugations are present even here (Russell et al. 1987), leading to a suggestion that they are produced by curvature forces pulling flux tubes draped over the ionopause into the ionosphere. Further from the subsolar ionopause, waves and flux ropes might be produced by the Kelvin-Helmholtz instability (Wolff et al. 1980). Bertucci et al. (2005) inferred the presence of ripples on the Martian magnetic pile up boundary at large solar zenith angles from discrepancies between model normals and those determined from minimum variance analysis. Although sharp ionopause density gradients are the norm at Venus, Duru et al. (2009) noted that they were only observed in 18% of the samples studied at Mars. Noting past work indicating highly fluctuating electron densities in the Martian ionosphere, Duru et al. (2009) attributed the infrequent occurrence of strong ionopause density gradients at Mars to time or spatially dependent phenomena, perhaps the Kelvin-Helmholtz instability at the ionopause. Pope et al. (2009) inferred the presence of giant vortices capable of redistributing and causing the substantial loss of ionospheric plasma at Venus.

Simulations of the Kelvin-Helmholtz instability at the ionopauses of the unmagnetized planets often reach conflicting conclusions. Terada et al. (2002) reported results from a two-dimensional hybrid code simulation which indicate preferential wave growth beginning even at the subsolar point and continuing further antisunward in the hemisphere with an electric field pointing away from the planet, as shown in Fig. 14. By contrast, the waves began to develop only at greater solar zenith angles in the opposite hemisphere. Penz et al. (2004) used an MHD simulation to study the case where flows lie transverse to draped magnetosheath magnetic field lines. Under these conditions, the magnetic field plays no role in stabilizing the instability. The subsolar magnetopause is stable and the non-linear instability develops on the equatorial flanks. They estimated that the atmospheric loss via the Kelvin-Helmholtz instability is comparable to that by other non-thermal loss mechanisms. Amerstorfer et al. (2007) reported results from an MHD simulation with a similar magnetic field configuration, this time in the presence of strong radial gradients in density and velocity at the terminator ionopause. High magnetosonic Mach numbers (increasing compressibility) and greater magnetosheath to ionospheric density ratios diminish the likelihood of the instability. The

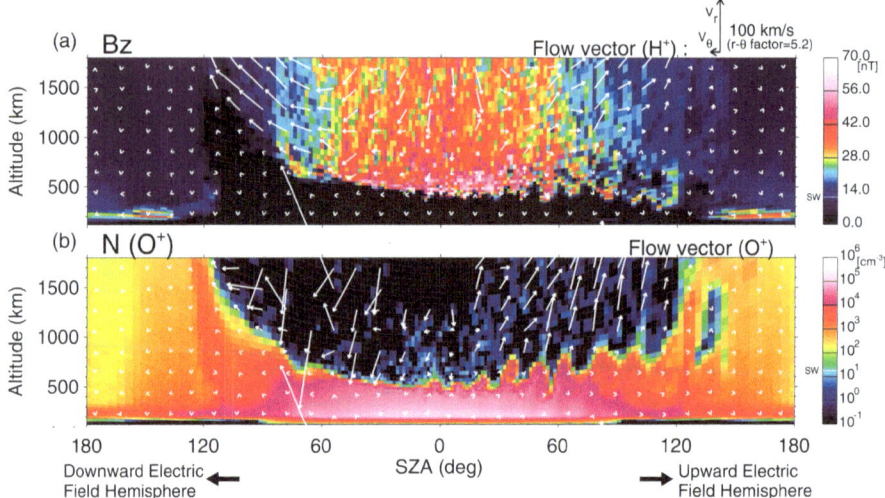

Fig. 14 Results from a global multifluid hybrid code simulation for the solar wind interaction with Venus as a function of solar zenith angle (SZA) and altitude (Terada et al. 2002). Panel (a) shows the distribution of magnetic field strengths (colors) while vectors show flow directions for plasma of solar wind origin. Panel (b) shows the densities and flow directions for O^+ ions of planetary origin. The ionopause exhibits much greater corrugation in the hemisphere with an upward electric field than the hemisphere with a downward electric field

wavelengths of the fastest growing mode diminish as the density ratios increase. MHD simulation results indicating a sharper ionopause and a greater shear flow led Bößwetter et al. (2007) to conclude that the Kelvin-Helmholtz instability is more likely in the hemisphere in which the convection electric field points towards the planet. Amerstorfer et al. (2010) then considered the evolution of the instability from a linear phase, through a nonlinear phase with regular structures through a turbulent phase with nonlinear structures. They concluded that the instability could account for the atmospheric loss rate estimated from observations. Finally, Möstl et al. (2011) used an MHD simulation to argue that conditions generally do not favor the occurrence of the Kelvin-Helmholtz instability at Venus, with the possible exception of the dayside induced "magnetopause", or upper boundary of the magnetic barrier, during solar maximum, at dayside locations away from the subsolar point, where magnetosheath flows lie transverse to draped magnetosheath magnetic field lines.

Global images have the potential to play a decisive role in testing the often conflicting predictions of the various solar wind-planetary interaction mechanisms proposed to occur at Mars and Venus. They can be used to determine occurrence patterns and extent as functions of solar wind and solar cycle conditions, and to quantify the importance of each mechanism to the loss of planetary atmospheres. As an example, take the case of the bow shock asymmetries expected in response to atmospheric loss via ion pick-up. Observations can first be used to test conflicting model predictions indicating that the bow shock lies further from (closer to) the planet in the hemisphere with outward pointing convection electric fields. The degree to which the bow shock is asymmetric can provide information concerning the significance of ion pick-up over time scales ranging from minutes to solar cycles. Similarly, observations of the size, extent, and number of the wavy density structures generated by the Kelvin-Helmholtz instability or with isolated structures associated with the flux ropes generated by magnetic reconnection can be used to determine the importance of these mechanisms as a function of simultaneously measured solar wind conditions.

2.4 The Moon

Despite its tenuous nature, the lunar exosphere remains high on the list of planetary science targets thanks to its complexity and role as an accessible representative of airless bodies in the solar system and the possible presence of water and other potential resources. The role of volatiles in the lunar exosphere is particularly important.

The solar wind and meteoroids deliver protons and other species to the lunar surface at local rates that depend on surface composition, impinging local topography, and the presence of structures such as magnetic anomalies. Solar wind ions weather the surface by creating defects in the lattice that weaken the solid state structure. Because the lunar surface is generally saturated with these volatiles, the implanted species escape the surface and form the volatile lunar exosphere through a variety of processes including sputtering, recoil, and diffusion. These processes deposit H and other volatiles into cold traps and form OH (and possibly water) through chemical alteration of oxygen-bearing minerals. Exospheric volatiles are reclaimed by the solar wind as picked-up photoions and charge-exchange products. Global imaging of the total lunar exosphere including all species at regional scales as functions of solar zenith angle and the plasma and space environment will lead to a unified understanding of the plasma, exospheric, and geologic Moon.

The Lunar Atmosphere Dust and Environment Explorer (LADEE, Elphic et al. 2014) Neutral Mass Spectrometer (NMS, Mahaffy et al. 2014) confirmed the presence of water in the equatorial lunar exosphere for brief periods early in the instrument turn on/warm up period. These detections of non-polar exospheric water occurred preferentially near the radiant of episodic meteor stream encounters. Water densities of 2–3×10^8 m^{-3} during these meteor shower events (Benna et al. 2015a) correlate nicely with LADEE Lunar Dust Experiment (LDEX, Horányi et al. 2014) dust stream occurrences.

Although LADEE established that lunar volatile gases like water can be released by the impact of solar system objects like meteoroids in the equatorial region, volatiles can also be released from the interior of the Moon, through moonquakes (Cook and Stern 2014). Additionally, they can be synthesized in the upper layer of the lunar regolith by the solar wind. Once released, they are transported across the lunar surface until they either escape to space or become trapped in cold permanently shadowed regions (PSRs) that have maintained temperatures below 100 K for billions of years. Of particular interest is water trapped in these PSRs.

In fact, there has been observational verification of an active water and hydroxyl environment (i.e., water cycle) at the Moon including Lunar Crater Observation and Sensing Satellite (LCROSS, Schultz et al. 2010) confirmation of water existing within the lunar polar cold traps (Colaprete et al. 2010; Schultz et al. 2010). Other evidence includes data from a set of IR sensors showing an OH veneer that extends all the way down to the lunar equator, and which may even possess a present-day, dynamic diurnal component (Pieters et al. 2009; Clarke et al. 2009; Sunshine et al. 2009). The distribution of water in the lunar polar regions is heterogeneous on all observed scales (Mitrofanov et al. 2010).

However prior to deposition into cold traps, the volatiles must be transported some distance across the lunar surface. Volatile mobility depends on many parameters including species, surface composition, and temperature. For example, the argon density distribution results from a surface interaction, an excess of adsorption over desorption on the nightside as the lunar surface cools, so its density peaks at the terminator where the surface heats up (Hodges 1977). Helium, on the other hand, is not adsorbed onto the surface so it spends more time on the cold nightside than on the warmer dayside because the lateral extent of its trajectories is proportional to temperature (Hodges 1973, 1975). Consequently, He density peaks on the nightside. Of course, the scale height and its dependence on temperature also play a role.

In general, the cold nightside lunar atmosphere is dominated by non-condensible species, including He, detected by Apollo-era instrumentation, and Ne and H2, as observed by LADEE and the Lunar Reconnaissance Orbiter (LRO, Tooley et al. 2010). LADEE also confirmed the presence of argon at the equator and the Lyman-Alpha Mapping Project (LAMP, Gladstone et al. 2010) placed limits on Ar at the poles (Hodges and Mahaffy 2016; Grava et al. 2015; Benna et al. 2015b). These *in situ* observations, when coupled with global data on the structure of the lunar exosphere including local time dependence and vertical scale heights, are essential for determining production rates and polar cold trapping efficiencies. With guidance from modeling efforts, global images of total exospheric content could determine the constituents of the lunar atmosphere over the poles.

Global imaging will also reveal the relationship between the time-variable solar wind flux and composition and the lunar exosphere. This would be accomplished in a manner similar to what LRO did *in situ* (Feldman et al. 2012) by showing that the surface He density exhibits variations responding to changes in the solar wind alpha flux (see also Benna et al. 2015b). Of course, global imaging would provide an overall perspective on this process not possible with local *in situ* measurements. Global images of the lunar atmosphere can also be used to study the behavior of the lunar exosphere as the Moon moves in and out of the terrestrial magnetotail, modulating solar wind sputtering and enabling identification of the distant terrestrial magnetopause and possibly bow shock. Imaging can also reveal the global effects of meteoroid bombardment (e.g., Colaprete et al. 2016).

Soft X-ray observations of the lunar exosphere complement and validate model predictions for the dominant contributors to the exospheric column density. Furthermore, because soft X-ray imaging relies on the presence of the solar wind, global imaging will reveal the shape and extent of structures effected by the solar wind-lunar interaction. Plasma structures in the vicinity of the Moon include a low density wake (Lyon et al. 1967; Zhang et al. 2014) and mini-magnetospheres above magnetic anomalies (Wieser et al. 2010) that could be imaged globally using soft X-ray emission. In addition to morphology, global imaging will reveal aspects of the interaction that can be quantitatively compared to model predictions, for example the extent over which solar wind ions impact the lunar surface beyond the terminator (Collier et al. 2014).

Global images will supply the key to a unified understanding of the plasma, exospheric, and geologic Moon. They will provide information on the exospheric content as a function of altitude and location above the lunar surface that can be correlated to geologic regions. Global imaging will also reveal properties of the solar wind plasma-lunar interaction, such as wake morphology and how magnetic anomalies affect solar wind implantation.

3 Soft X-Ray Intensities from Solar Wind Charge Exchange

Solar wind charge exchange is responsible for EUV and soft X-ray emission not only in regions of the solar system where the solar wind interacts with neutral gases from objects such as comets, and planetary exospheres such as Earth's geocorona, but also with the interstellar gas flowing through the heliosphere (see Sect. 4.2). Charge exchange leaves the product ion in an excited state which then returns to the ground state through the emission of one or more photons. Since the bulk of the ions in the solar wind are highly ionized, most of these photons are in the soft X-ray and extreme ultraviolet. As shown in Fig. 15, the resulting spectrum is extremely rich. However, this spectrum is generally observed at relatively low

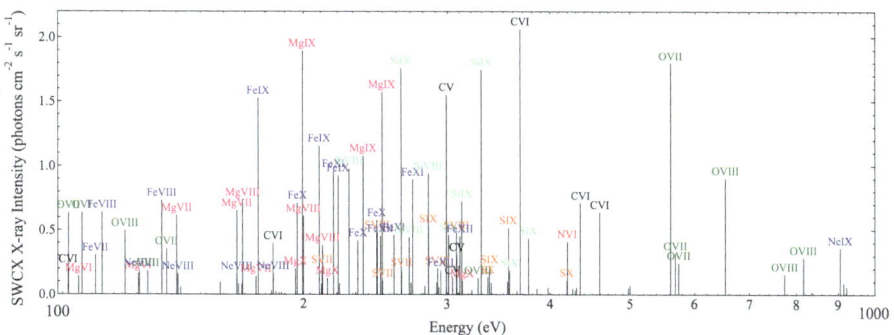

Fig. 15 Model solar wind charge exchange spectrum (similar to Fig. 1 from Koutroumpa et al. 2009b)

spectral resolution so that the bulk of the lines are severely blended. This section reviews the production of EUV and soft X-rays by solar wind charge exchange. In particular it describes the many factors required to determine the spectrum seen by an observer looking along a single LOS. Calculating the spectrum involves many quantities that are poorly known. However, we will show that working at lower resolution makes the problem in some ways more tractable.

The intensity, I (in photon cm^{-2} s^{-1}), of the emissions from transition j, for species s, in charge state q, seen by an observer looking along a LOS is given by the integral through the emitting region(s) along that line of sight:

$$I_j = \int P_{sqj} dl = \sum_n \int n_n n_q v_{rel} \sigma_{sqn} b_{sqj} d\Omega dl / 4\pi, \tag{1}$$

where P_{sqj} is the volume emission rate (photons cm^{-3} s^{-1}) for a specific transition j (with photon energy E_j) from a specific charge exchange collision of the solar wind ion species (denoted s) in charge state q with a neutral target n. The summation over n reflects the reality that there may be multiple neutral species, though in many cases we need only consider H and He (for the diffuse heliospheric emission) or only H (for the Earth's magnetospheric emission). The relevant charge exchange cross section at the appropriate collision energy is σ_{sqn}, and b_{sqj} is a branching ratio in the product ion species for the transition of interest. The branching ratio is the fraction of ions undergoing charge exchange between n and sq that relax through transition j. The densities n_n and n_q are those of the neutrals and ions respectively, while v_{rel} is the relative velocity of the neutrals and ions:

$$v_{rel} \sim \left(v_r^2 + v_{therm}^2\right)^{\frac{1}{2}} \tag{2}$$

where v_r is the bulk velocity of the ions, it being supposed that the thermal velocity of the target neutrals is small. The quantity v_{therm} is $3kT/m_p$ for the solar wind ions. The photon flux within some field of view is the integral of intensity over solid angle increment $d\Omega$. This equation is generally rewritten as:

$$I_j = \int P_{sqj} dl = \int n_n n_p v_{rel} \frac{n_q n_s}{n_s n_p} \sigma_{sqn} b_{sqj} d\Omega dl / 4\pi, \tag{3}$$

where n_p is the solar wind proton density. To calculate the integral intensities along specific lines of sight, each of these factors must be considered in detail.

It should be noted that in many cases, instrumental resolution is insufficient to isolate individual lines. For a bandpass containing emission from many different transitions (j) from many ion species (sq) charge exchanging with different neutral targets (n)

$$I = \sum_j \sum_{sq} \sum_n \int n_n n_p v_{rel} \frac{n_q n_s}{n_s n_p} \sigma_{sqn} b_{sqj} \, d\Omega \, dl / 4\pi \qquad (4)$$

which can be rewritten as

$$I = d\Omega / 4\pi \sum_n \int n_n n_p v_{rel} dl \left[\sum_j \sum_{sq} \frac{n_q n_s}{n_s n_p} \sigma_{sqn} b_{sqj} \right] \equiv \sum_n \frac{d\Omega}{4\pi} Q_n \varsigma_n \qquad (5)$$

where

$$Q_n \equiv \int n_n n_p v_{rel} dl \quad \text{and} \quad \varsigma_n \equiv \sum_j \sum_{sq} \frac{n_q n_s}{n_s n_p} \sigma_{sqn} b_{sqj}. \qquad (6)$$

Here we have assumed that n_q/n_s, n_s/n_p, and σ_{sqn} (which must be a function of v_{rel}) are at least relatively constant along the line of sight. This formulation segregates the bulk properties of the solar wind and its neutral targets from the atomic data.

An alternate formulation seen in the literature for both individual lines and band passes is more convenient for calculating energy fluxes (in $eV\,cm^{-2}\,s^{-1}$):

$$F_j = \int E_j P_{sqj} dl = Q\alpha_j \qquad (7)$$

where

$$\alpha_j \equiv E_j \frac{n_q n_s}{n_s n_p} \sigma_{sqn} b_{sqj} \qquad (8)$$

that is, the energy-weighted cross section, which is also called an *emission scale factor*.

As we will see below, Q can be derived from models with some reasonable degree of confidence, while ς, a *production factor*, requires atomic data that is, in many cases, unknown. In Sect. 3.1.3 we will demonstrate that ς has been derived from observations for the broad Röntgensatellit (ROSAT, Trümper 1992) $\frac{1}{4}$ keV bandpass, which allows simulations of the entire magnetosheath in that and similar bands.

3.1 Theoretical and Observation-Inferred Charge-Exchange Cross Sections

3.1.1 Predicted Charge-Exchange Cross-Sections

The collision of an ion with a neutral target can result in the transfer of an electron from the neutral atom or molecule to the ion, i.e., charge exchange. The incident ion with charge, q, is represented as M^{q+}, where M is a minor species such as O, N, C, and Fe. Of greatest interest are high charge state ion species such as O^{7+} or C^{6+} that are abundant in the solar wind and produce X-rays upon recombination. The target neutral species is designated B, where B can be H_2O for comets, H for the Earth's exosphere, H and He for interplanetary space, or other neutral species as required. The relevant charge exchange reaction can be written:

$$M^{q+} + B \rightarrow M^{(q-1)+} + B^+ \qquad (9)$$

Fig. 16 Interaction of potential energy curves versus inter-nuclear separation for two atomic species undergoing a charge exchange collision. (Adapted from Isler 1994, see text for additional information)

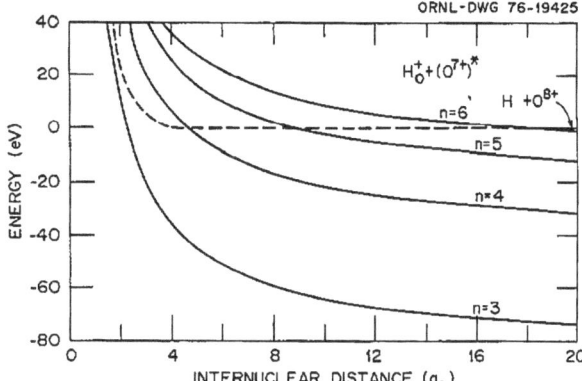

A key example of this reaction in the terrestrial exosphere or in the heliosphere is:

$$\dot{O}^{7+} + H \rightarrow O^{6+} + H^+ \tag{10}$$

For highly ionized recipients (i.e., large values of q), the product ion species (i.e., $M^{(q-1)+}$) is invariably left in a highly excited state such that a radiative cascade follows the collision as the excited ion de-excites to the ground-state. For large values of q at least one of the photons in this cascade is an EUV or a soft X-ray photon.

At solar wind energies (i.e., ~ 1 keV/amu), the cross section σ_{sqn} for this type of charge exchange collision is very large, much greater than geometrical based on the dimensions of the interacting particles. Figure 16 shows interaction potential energy curves versus inter-nuclear separation for two atomic species undergoing a charge exchange collision (Isler 1994). The dashed line indicates results for incident reactants O^{8+} and H, which experience a weak point charge-induced dipole interaction whose energies remain almost constant at larger distances. By contrast, the outgoing reaction products O^{7+} and H^+ experience a strong Coulomb interaction. The curves include the energies of the hydrogen-like O^{7+} excited to states with principle quantum number n. If there is a suitable resonance for the electron in the system, then the charge exchange reaction can be said to take place with some probability at a curve crossing. The curve crossings for high values of n take place at large radii (e.g., $r \approx 9a_0$ for $n \approx 5$, where $a_0 = 0.54 \times 10^{-8}$ cm is the Bohr radius). Hence, for a reaction probability of about 0.5 the cross section would be $\sigma \approx \pi \times (9a_0)^2 \approx 5 \times 10^{-15}$ cm^2.

The probability of charge-exchange depends not only on the principal quantum number, but on all of the values that describe the state into which the electron is initially inserted. The cross-sections, σ_{sqn}, referred to above are the *total* cross-section over all possible initial states, while the details depending on the initial states are hidden in the b_{sqj}. Calculated cross-sections must be constructed from those of each initial state. Total cross-sections can be measured, but n, l, m resolved cross-sections produce far more insight into the physical process.

Figure 17 demonstrates the resonance process that necessitates the energy of the final state of the recipient ion product in the charge exchange reaction. It shows the electron potential energy as function of distance along the inter-nuclear axis for (in this case) the $B^{5+} + H \rightarrow B^{4+}(n = 3) + H^+$ system at a time during the reaction when the nuclei are 9 atomic units (a.u.) apart (Cravens 2002). Here 1 a.u. of distance is 1 a_0 and 1 a.u. of energy is 1 Rydberg or 27.2 eV. This internuclear distance is a favorable one because the electron energy can remain about the same (the resonance) and "move over" from the region near the H^+ nucleus over a low energy barrier to the region near the B^{x+} nucleus.

Fig. 17 Electron potential energy as a function of distance along the inter-nuclear axis for the $B^{5+} + H \rightarrow$ $B^{4+} (n = 3) + H^+$ system at a time during the reaction when the nuclei are 9 atomic units (a.u.) apart (adapted from Cravens 2002)

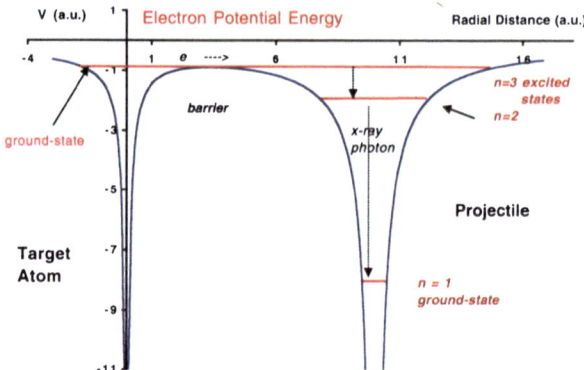

The simple classical over-barrier (COB) collision model provides approximate cross sections and excitation levels (Mann et al. 1981). The cross sections are energy-independent and apply only to relatively low collision energies but are appropriate for solar wind ions. For a fully stripped recipient species M^{q+} colliding with a neutral target species B that has an ionization potential of I_p, the energy defect is $\Delta E = q^2/2n^2 - I_p$ and the curve crossing distance is $R_x \approx (q-1)/\Delta E$, where $I_p = 0.5$ a.u. for atomic H. The COB cross section is then $\sigma \approx \pi R_x^2$ and the most likely excitation level (i.e., the principle quantum number) is given by $n \leq q\{2I_p[1 + (q-1)/(2q^{\frac{1}{2}} + 1)]\}^{-\frac{1}{2}}$. For the $O^{8+} + H$ collision, this gives $n \approx 5$ for the product O^{7+} and $\sigma \approx 250\, a_0^2 \approx 7 \times 10^{-15}$ cm^2.

3.1.2 Experimental Measurements of Charge-Exchange Cross Sections

Numerous laboratory measurements of high charge state ion collisions with neutrals have been made over the years. Gilbody (1986) reviewed some of the earlier experimental work, reporting for example on laboratory measurements and theoretical calculations for $C^{6+} + H$ charge exchange cross sections as a function of energy. In this case the measured charge exchange cross section is 3×10^{-15} cm^2 for a collision energy of 1 keV/amu, which greatly exceeds the geometrical cross section. Janev and Winter (1985), and Janev et al. (1983, 1988) reported measurements of state-selective cross sections that indicated the product ion is left excited with a high principle quantum number (e.g., $n \approx 4$).

The cross sections σ_{sqn} are velocity dependent, but approximately constant for most solar wind species as a function of the relative velocity between the interacting particles over reasonable velocity ranges. While some important emission lines like O^{6+} and O^{7+} have been reasonably well characterized, many others that contribute to the emissions with energies < 500 eV that are relevant to imaging solar wind interactions with the planetary objects are not (see Smith et al. 2014, for a discussion of alternative methods).

Beiersdorfer et al. (1999, 2000, 2001, 2003), Wargelin et al. (2008), Greenwood et al. (2001), Mawhorter et al. (2007), and Betancourt-Martinez et al. (2014) present some relatively recent experimental measurements of X-ray emissions generated by charge exchange. (See also the review within Krasnopolsky et al. 2004.) The more recent experiments include a wide variety of target (i.e., H_2O and CO_2) and incident ion species and charge states. A variety of experimental methods and incident energies were employed. For example, Greenwood et al. (2001) measured the X-ray spectrum emitted during the charge exchange process using a crossed-beam experiment in addition to determining the initial and final charge states of the recipient ions.

Fig. 18 Energy level diagram for O^{6+} (O^{6+} X-ray emission lines)

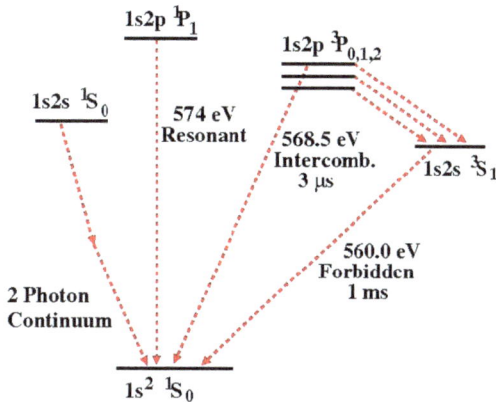

Beiersdorfer et al. (2003) used a microcalorimeter to measure the X-ray spectrum generated when trapped ions interact with neutral targets in the Lawrence Livermore electron beam ion trap (EBIT-I). For O^{7+} ions interacting with neutrals like CO_2, the resulting helium-like emission spectrum from O^{6+} produces X-ray transitions such as those observed at comets with especially strong lines near 570 eV. In particular, the microcalorimeter detected strong emission from the forbidden transition $1s2s\ ^3S_1$–$1s^2\ ^1S_0$ at 564 eV. Figure 18 shows an energy level diagram for O^{6+} illustrating the forbidden, resonant, and intercombination lines for the $n = 2$ to $n = 1$ transition.

Employing ACE measurements of the solar wind composition, Whittaker et al. (2016) calculated the charge exchange emission scale factors for O^{7+} and O^{8+}. They showed that the scale factors peak sharply near 8.2×10^{-17} eV cm^2 at solar wind velocities of ~ 400 km s^{-1}, diminishing rapidly for lower velocities and more gradually for higher velocities.

Values used for the emission scale factor for the X-ray band with $E > 50$ eV range from 6×10^{-16} eV cm^2 (Cravens et al. 2001; Robertson and Cravens 2003b) to 9.4×10^{-16} eV cm^2 (Pepino et al. 2004) to 1.5×10^{-15} eV cm^2 (Robertson and Cravens 2003b; Cravens et al. 2001) for the slow solar wind (all of whom cite Schwadron and Cravens 2000 in one way or another for their values) and 3.3×10^{-16} eV cm^2 for the fast solar wind (Pepino et al. 2004).

Summarizing, the charge exchange cross sections for many interactions that generate emissions in the 0.1–0.284 keV band remain both poorly understood and poorly determined. Theory and observations indicate a wide range of values for charge exchange cross-sections and their corresponding emission scale factors for H interacting with O^{7+} and O^{8+}. As we shall see (Sects. 5 and 6), the band-integratedth cross-section for solar wind charge exchange is best estimated from well calibrated observations in space.

3.1.3 A Band-Averaged Production Factor for the 1/4 keV Band

As will be described more fully in Sect. 6.1, $\frac{1}{4}$ keV X-ray emission from the magnetosphere was observed by ROSAT as a temporally variable background component which was measured and removed from the ROSAT All-Sky Survey. The temporally variable emission was later shown to be well correlated with the solar wind flux. The relation between observed

X-ray emission and the solar wind flux is

$$\text{ROSAT counts s}^{-1}\text{ degree}^{-2} = (0.083 \pm 2.26) \times 10^{-2} + (0.186 \pm 0.009)\left[\frac{n_{sw}v_{sw}}{10^8}\right] \quad (11)$$

where $n_{sw}v_{sw}$ is in cm^{-2} s^{-1} (Kuntz et al. 2015). Given the $Q \equiv \int n_n n_p v_{rel} dl$ through which ROSAT observed, one can calculate the $\frac{1}{4}$ keV band-averaged production factor. Modeling the magnetosheath for each observation of the All-Sky Survey is computationally prohibitive. Kuntz et al. (2015) used a suite of extant MHD runs to do the equivalent; given the ROSAT observing geometry, they determined the typical Q through which ROSAT would have observed as a function of the solar wind flux.

$$\left[\frac{n_{sw}v_{sw}}{10^8}\right] = (0.037 \pm 0.189) + (20.68 + 0.41)\left[\frac{Q}{10^{20}}\right] \quad (12)$$

where Q is in cm^{-4} s^{-1}. The combination of these two relations yields

$$\text{ROSAT counts s}^{-1}\text{ degree}^{-2} = (3.86 \pm 0.20) \times 10^{-20} Q. \quad (13)$$

We can then use this production factor to determine the ROSAT $\frac{1}{4}$ keV count rate for a given Q and, given a model for the shape of the spectrum of the emission, can convert this to the count rate for any instrument with a similar band-pass.

Using the available atomic data rather than ROSAT observations, Robertson et al. (2009b) calculated an equivalent production factor for the slow solar wind interacting with neutral H: 8.51×10^{-21} count cm^4 deg^{-2} which is a factor of 4.53 smaller than the value derived by Kuntz et al. (2015) above. Koutroumpa et al. (2009a), also using a collection of atomic data, created a solar wind charge exchange spectrum for a slow solar wind interacting with neutral H. The Koutroumpa et al. (2009a) spectrum contains 2.5 more flux than the Robertson et al. (2009b) spectrum, and thus would have a production factor of 2.13×10^{-20} count cm^4 deg^{-2}, only a factor of 1.79 lower than the Kuntz et al. (2015) measurement. Since the atomic data used is uncertain and likely to be missing many of the fainter transitions, this agreement between model and measurement is surprisingly good.

3.2 The Branching Ratio and Spectra

Once the charge transfer collision has occurred and the product ion is produced with a high principle quantum number, n, and angular momentum quantum number, l, a radiative cascade ensues, subject to the relevant selection rules (e.g., $\Delta l = \pm 1$ for dipole-allowed transitions), so that the ion eventually ends up in the ground-state.

The details of this cascade depend on the set of radiative transition probabilities (or Einstein A coefficients), and are encompassed in the branching ratio coefficient, b_{sqj}, which appears in the intensity expression given above (see Eqs. (1) and (3)). The coefficient in this equation is an average that must also include information on the initial quantum numbers of the product ion. For example, if an O^{5+} is created in the $n = 4$ and $l = 1$ state (2s4p) by charge exchange of O^{6+} with a neutral, then it can radiate to the following states: $2s^2$ with fraction 0.77, 2s3s with fraction 0.11, and 2s3d with fraction 0.04. The 2s2p state will then radiate 100% to the ground state $2s^2$. The difficult task of finding b_{sqj} for all species and charge states relevant to solar wind charge exchange remains only partially and approximately complete.

Fig. 19 The total emission spectra from the single (dashed lines) and sequential (solid lines) collision regions of a cometary atmosphere normalized to unit flux of solar wind (SW) ions. Collisions of C^{4+}, C^{5+}, C^{6+}, N^{7+}, O^{6+}, O^{7+}, O^{8+}, and Ne^{8+} with cometary neutral atoms and molecular constituents are considered. The energy resolution of the discrete spectral lines Γ is taken arbitrarily as 1 eV (adapted from Kharchenko and Dalgarno 2000)

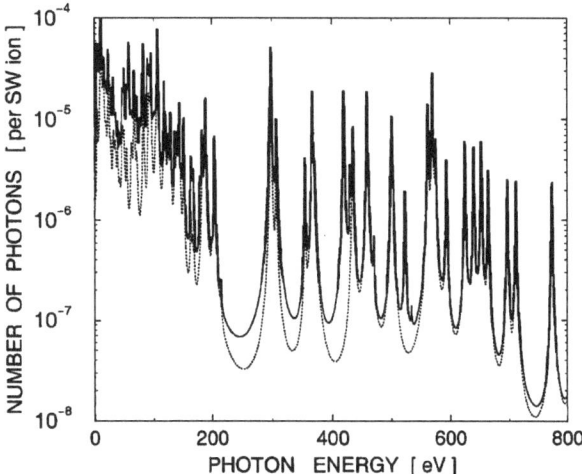

For the solar wind ions of interest here, some of the resulting transitions are in the EUV and soft X-ray parts of the spectrum. For example, O^{7+} ions (produced by charge exchange from solar wind O^{8+} ions) generate a hydrogen-like spectrum (O^{7+} emission lines in X-ray astronomy notation) while O^{6+} ions (from O^{7+}) generate a helium-like spectrum (O^{6+} emission lines), as discussed earlier. These lines are in the soft X-ray part of the spectrum. Other recipient species with different charge states produce different spectra.

The detailed X-ray spectrum resulting from solar wind charge exchange (SWCX) depends on the abundances (or fluxes) of the highly-charged heavy ions in the solar wind (e.g., C^{6+}, O^{7+}, O^{8+}, Mg^{12+}, Fe^{13+}, etc.). These abundances depend on where on the Sun the solar wind originated, as discussed by von Steiger et al. (2000), and Schwadron and Cravens (2000). For example, the slow (300 km s^{-1}) solar wind originates from a hotter solar corona and has a relatively higher O^{7+}/O^{6+} ratio than the fast (700 km s^{-1}) solar wind that originates from cooler parts of the solar corona. Solar wind composition is discussed in more detail in the next section.

Given the solar wind heavy ion abundances and relevant charge exchange cross sections and radiative cascade probabilities, both the EUV and soft X-ray spectra and the efficiencies of the SWCX mechanism can be determined. Several authors have undertaken this exercise both for detailed spectra and for broad-band X-ray emission bands with and without instrumental response functions included for specific observations such as those made by ROSAT (cf., Kharchenko and Dalgarno 2001; Pepino et al. 2004; Krasnopolsky et al. 2004; Robertson et al. 2009b). Figure 19 shows a cometary SWCX spectrum calculated by Kharchenko and Dalgarno (2000). The strong O^{6+} lines near 570 eV are particularly obvious but a large number of other lines are present at lower energies.

3.3 The Flux of High Charge State Solar Wind Ions

We take the flux (F_{sq}) of ions of species s with charge state q to be proportional to that of protons, i.e., $F_{sq} \approx f_{sq} n_{sw} u_{sw}$, where f_{sq} is the relative abundance of individual heavy ion species s in charge state q, n_{sw} is the proton density and u_{sw} is solar wind proton velocity. Consistent with this assumption, we note that Neugebauer et al. (2000) reported a close correlation between OMNIWeb proton and SOHO O fluxes over a period of 8 days during 1996. Whittaker et al. (2016) found that simulated and observed integrated line-of-sight

Fig. 20 The SWICS O^{+6} flux *versus* the SWEPAM proton flux from the ACE data binned to 1 day intervals covering 1998 to 2011. The *blue* line is the canonical O/H value of 0.00045. The *red* boxes are the means for $\delta \log(n_p v_p) = 0.1$ bins, while the bars show the dispersion $[\sum(x^2)/n - (\sum(x)/n)^2]^{0.5}$ in each bin

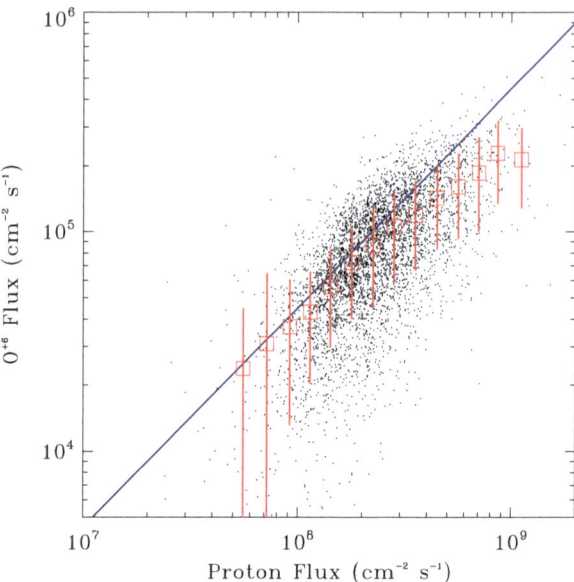

Table 4 Solar wind properties (Ebert et al. 2009)

Solar wind	Density (cm^{-3})		Bulk speed (km s^{-1})		Thermal speed (km s^{-1})
Type	H	He	H	He	H
Slow	5.55	0.13	392	399	45
Fast	2.12	0.11	745	769	79
ICME	5.9	0.32	449	450	51

charge exchange soft X-ray intensities agree better when the ratio of oxygen to proton flux in the model is held constant than when it is allowed to track that observed simultaneously in the solar wind. Figure 20 shows ACE SWICS (Gloeckler et al. 1998) observations of the flux of O^{6+} ions versus ACE SWEPAM (McComas et al. 1998) observations of the flux of protons from 1998 to 2011. The plot exhibits considerable scatter, but the two fluxes are well correlated, confirming the assumption of a relatively constant ratio.

Nevertheless, the relative heavy ion abundances differ for the slow and fast solar wind flows, with the slow components originating at equatorial latitudes and the fast components originating at coronal holes. Table 4 summarizes average slow solar wind proton and He densities, bulk speeds, and thermal speeds at Earth. Near the ecliptic plane, 90% of slow-stream density measurements lie between 1 and 20 cm^{-3}. von Steiger et al. (2013) report H/O ratios of 1500, which gives total O densities of 2.0×10^{-3} cm^{-3} for proton densities of 3 cm^{-3}. The slow solar wind composition is very biased towards elements with low first ionization potentials and matches that of the solar corona (Schwadron et al. 1999; von Steiger et al. 2000). Predominant species are: C^{5+}, C^{6+}, O^{6+}, and O^{7+} (Schwadron and Cravens 2000; von Steiger et al. 2000).

Table 4 also summarizes typical fast solar wind proton and He densities, bulk speeds, and thermal speeds. Near the ecliptic plane, 90% of fast-stream H density measurements lie between 1 and 10 cm^{-3}. Corresponding average high charge state O densities are slightly

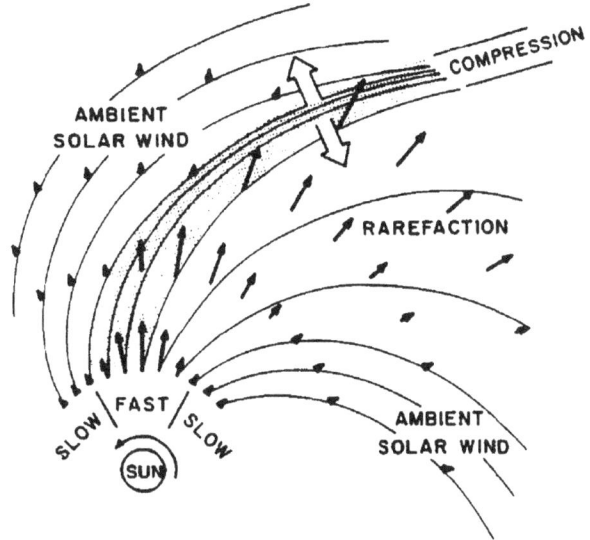

Fig. 21 Fast and slow solar wind stream structure (Pizzo 1978)

greater than those in the slow solar wind, $\sim 2.2 \times 10^{-3}$ cm^{-3}. The fast solar wind composition exhibits less of a bias towards elements with low first ionization potentials and matches that of the photosphere. Predominant species are: C^{5+}, N^{5+}, O^{6+}, and Ne^{8+}.

As illustrated in Fig. 21, fast solar wind streams overtake slow solar wind streams, creating compressional corotating interplanetary regions (CIRs, Pizzo 1978) with enhanced densities and magnetic field strengths that spiral outward from the Sun. Fast-moving, outward-propagating, interplanetary coronal mass ejections (ICMEs) occasionally disrupt this two-stream pattern, particularly at the peak and during the declining phase of the solar cycle. The azimuthal extents of ICMEs at Earth are about four times greater than their ~ 0.1 AU half-widths (Russell and Mulligan 2002). Typical ICME proton and He densities, bulk speeds, and thermal speeds are again listed in Table 4 (Ebert et al. 2009). Near the ecliptic plane, 90% of ICME density measurements lie between 1 and 10 cm^{-3}. The overall chemical composition of ICMEs resembles that of the slow solar wind, but the He density is frequently enhanced, with the ratio of He to proton densities often exceeding 8% and sometimes exceeding 10%. Minor ions with $Z > 2$ in CMEs have enhanced densities with respect to protons. This has been particularly well-documented in the Fe/H ratios (e.g., Ipavich et al. 1986; Mitchell et al. 1983; Bame et al. 1979). Furthermore, the ionization temperatures inferred from the charge state distributions are also frequently, but not always, elevated in comparison to coronal hole flow, which is also high-speed, and interstream flow. For example, coronal hole oxygen charge states indicate a 1.3 MK ionization temperature whereas CME-related charge states indicate an ionization temperature above 2 MK (Galvin 1997). Shocks propagating ahead of the ICMEs compress the local plasma, enhancing densities and magnetic field strengths.

The limited observations available indicate that the composition of high charge state heavy ion populations in the magnetosheath faithfully reflect those of their parent populations in the solar wind at much higher densities (Gloeckler et al. 1986). In the absence of any proposed process that would preferentially remove or accelerate high charge state heavy ion populations crossing the magnetopause, the same must be true for the ion populations in the LLBL, cusps, and mantle. Whether or not their differing Larmor radii enable ions with different mass to charge state ratios to reach different locations remains unknown.

Finally, the flux of ions F_{sq} can be written as $n_{sq}V_{rel}$, where n_{sq} is the density of species s with charge state q and V_{rel} is the effective velocity. Since charge exchange continues in hot plasmas like the magnetosheath and cusps with significant thermal velocities V_{th} even when the bulk velocity V_{bulk} vanishes, V_{rel} can be defined as $(V_{th}^2 + V_{bulk}^2)^{\frac{1}{2}}$.

3.4 Neutral Densities in the Outer Exosphere

Earth's exosphere is primarily H at all radial distances of interest here. It pervades all of the regions mentioned above, including the magnetosphere, cusps, boundary layers, the magnetosheath, and the near-Earth solar wind. Chamberlain (1963) presented a model for the exosphere in which there is a transition with radial distance from atmospheric densities dominated by orbiting particles to densities dominated by ballistic and escaping particles, to one dominated solely by escaping particles. In the latter region, temperatures remain constant with increasing radial distance at values $\sim 12\%$ of those at the critical level deeper within the atmosphere at the altitude where collisions first become negligible. Hodges (1994) presented the results of a Monte Carlo simulation for the seasonal and solar cycle variation of exospheric densities at distances from Earth ranging from the exobase to $10\,R_E$ that included atoms on ballistic trajectories and collisions. The model does not include the enhanced hot exospheric neutral densities in the vicinity of the cusps and magnetopause that result from the charge exchange of energetic protons and exospheric hydrogen. Because exospheric temperatures are determined by the relatively cold values at the critical level, the effective temperatures governing the interaction of exospheric neutrals and geospace plasmas are determined almost exclusively by the thermal and bulk velocities of the geospace plasmas.

By contrast to the sharp plasma density gradients predicted by models for the solar wind-magnetosphere interaction, exospheric models predict relatively smooth and gradual transitions in neutral density. Hodges' exospheric model predicts that neutral densities at $10\,R_E$ peak near the equator during the equinoxes, but at off-equatorial latitudes during the solstices. They fall off approximately as R^{-3}, where R is the radial distance from Earth. Dayside values at $10\,R_E$ near local noon increase from > 22.5 cm^{-3} at solar wind radio fluxes at 2800 MHz, F10.7 $= 230$ to > 37.5 cm^{-3} for F10.7 $= 80$ during the equinoxes. Dayside values at $10\,R_E$ near local noon increase from > 27.5 for F10.7 $= 230$ to > 37.5 cm^{-3} for F10.7 $= 80$ during the solstices. Consistent with the predictions of this theoretical model, Zoennchen et al. (2011) inferred exospheric neutral densities from TWINS observations of scattered solar Lyα finding a gradual transition from greater nightside (~ 45 cm^{-3}) than dayside (~ 22.5 cm^{-3}) neutral densities at $10\,R_E$ from Earth during solar minimum. Nevertheless, they noted the possibility of 100% errors in the model at these distances from Earth. Figure 22 compares their results with those from previous empirical studies. Finally, Fuselier et al. (2010) presented a case study of energetic neutral atom observations from which they inferred a neutral density of only 8 cm^{-3} at the subsolar point. Given the uncertainties inherent in both the models and observations, neutral densities at $10\,R_E$ may be far greater than any of those listed above. Determining neutral densities at large geocentric distances is difficult because interplanetary Lyα glow intensities exceed those for geocoronal emissions beyond 8–10 R_E (Bailey and Gruntman 2011).

Recent TWINS observations suggest that the exospheric density may also vary as a function of geomagnetic activity. Bailey and Gruntman (2011) reported ~ 10–20% increases in the neutral H density at the onset of geomagnetic storms, with the magnitude of the enhancement scaling to the strength of the geomagnetic storm as measured by the minimum in the Dst index. Since the increases in neutral density last for periods on the order of a day or less, the likely source is additional particles on ballistic trajectories with lifetimes on the order of 13–18 hours.

Fig. 22 Comparison of radial H density profiles from recent analytic models (adapted from Zoennchen et al. 2013)

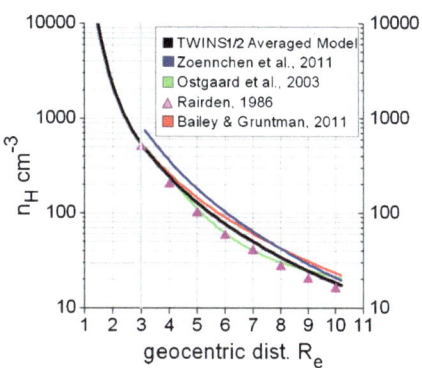

3.5 Charge Exchange Within the Magnetosphere

Ring current, Van Allen radiation belt, and plasmaspheric plasmas lie deep within the magnetosphere and exosphere, and are therefore also subject to charge exchange. Although plumes of cold (1–10 eV) plasmaspheric plasma can extend outward to the magnetopause with densities on the order of 10 cm^{-3} from locations deeper within the magnetosphere where densities are on the order of 1000 cm^{-3}, the plasmasphere often terminates abruptly at a sharp plasmapause that lies some 3–5 R$_E$ from Earth (Carpenter and Anderson 1992). Some plasmaspheric plasma flows upward through the cusps, joining the population of reflected solar wind ions entering the mantle and magnetotail. A ring current of hot (1–400 keV) plasma with densities ranging from 1–10 cm^{-3} encircles the Earth at radial distances from 2–7 R$_E$. Fluxes of energetic (> 50 MeV) ions peak in the inner Van Allen Radiation belt at distances of 1.2–3 R$_E$ from the center of the Earth, while fluxes of 0.1–10 MeV energetic electrons peak in the outer Van Allen Radiation belts at radial distances of 4–5 R$_E$ from Earth. Radiation belt ion densities are on the order of 1 cm^{-3} (Baumjohann and Treumann 1996; Hultqvist et al. 1999).

Since only high charge state heavy ion populations emit soft X-rays when they exchange electrons with neutrals, we must now consider the composition of the plasmasphere, ring current, and radiation belts to determine whether they are significant sources of soft X-rays. The plasmasphere is comprised of singly charged protons (93–97%), He (2–6%), and O (1%) ions (Moldwin 1997). The ring current is comprised of protons and, particularly during disturbed geomagnetic storm intervals, singly-charged O ions. Theory predicts that singly-charged ions dominate the radiation belt population, even for solar wind source species with purely high charge state populations (Spjeldvik and Fritz 1978). Although high charge state C and O ions may predominate at high energies (Cohen et al. 2017), particularly during injections (D. Mitchell, personal communication, 2017; Sibeck et al. 1988; Allen et al. 2017) the densities of these ions in the outer magnetosphere are insignificant, on the order of only 10^{-6}–10^{-3} cm^{-3} (Christon et al. 1994; Allen et al. 2016a,b, 2017). We conclude that magnetospheric particle populations are not a significant source of soft X-rays.

4 Other Sources of Soft X-Rays

The soft X-ray emissions generated by charge exchange with solar wind ions at Venus, Mars, Earth, the Moon, and comets must be distinguished from those generated by charge exchange with solar wind ions in the interplanetary medium, those generated by charge exchange with ions not of solar wind origin at Jupiter, those which result from a host of processes other than charge exchange at the planets (e.g., Bhardwaj et al. 2007), and the cosmic soft X-ray background. This section reviews the other sources that may lie in the line of sight of a soft X-ray telescope. With the aid of this information, we will proceed in Sects. 5 and 7 to predict the images that a wide field-of-view soft X-ray telescope would observe for confirmation by observations in Sect. 6.

4.1 Solar Emissions

The Sun is the brightest source of soft X-rays in the heliosphere. Its thermal plasmas generate both continuum and line emissions. Bremsstrahlung (or "braking radiation") represents a major source of continuum emission for hot ($\sim 10^6$ K) plasmas such as those found in the Sun's corona. Brehmstrahlung radiation is generated via the acceleration of charged particles colliding with targets such as atomic nuclei. Ion-electron recombination also results in emissions. In equilibrium plasmas, the ionization that results from the predominant electron-ion collisions is balanced by radiative and dielectronic electron-ion recombination. Both types of recombination generate line emissions whose energy depends on ion charge states and therefore on the ambient plasma temperature. For example, ion species such as O^{3+} are present in 10^5 K plasmas whereas O^{7+} is present in 10^6 K plasmas. The line radiation resulting from recombination lies mainly in the EUV for 10^5 K gases, but in the soft X-ray part of the spectrum for 10^6 K gases.

4.2 Emissions from the Heliosphere

Interstellar neutrals cross the boundaries of the heliosphere, enter the solar system, and exchange charges with solar wind ions that then generate soft X-rays. Due to the motion of the Sun through the local interstellar cloud, interstellar neutral H ($\sim 85\%$ by composition) and He ($\sim 15\%$) move with an apparent speed of about 26 km s^{-1} relative to observers in the solar system reference frame. Helium atoms appear to flow from an ecliptic longitude and latitude (λ, β) of $(255°, 5.5°)$, whilst H atoms appear to originate from a slightly different direction $(252°, 9°)$ (Lallement et al. 2005). The difference results from hydrogen neutrals exchanging charges with shocked protons at the distorted heliospheric interface, thereby forming a secondary neutral H population with the characteristics of the compressed protons.

As they move towards the inner solar system, neutral interstellar H and He atoms experience the Sun's effects differently. Neutral H atoms move sunward with the relative motion of the Sun and the Local Cloud and are affected by the attractive force of gravity and repulsive force of radiation pressure. Charge exchange with outward moving solar wind ions results in antisunward-moving neutrals and pick-up ions. Together with solar EUV ionization and electron impact ionization, this charge exchange excludes neutral H from a cavity around the Sun whose ~ 1–2 AU size depends on solar activity through the strength of the depletion processes (Quémerais et al. 2006).

The situation is very different for neutral He. Because radiation pressure has little effect on neutral He, these atoms execute Keplerian hyperbolic orbits to form the He focusing cone downstream from the Sun. EUV solar photons ionize the He atoms, but the resulting ionization cavity extends only about 0.5 AU from the Sun. Consequently, the Earth and spacecraft monitoring the solar wind at the L1 libration point pass through the substantially enhanced neutral He densities within the focusing cone once each northern hemisphere winter (e.g., Dalaudier et al. 1984; Gruntman 1994; Gloeckler et al. 2004). Just as in the case of the H

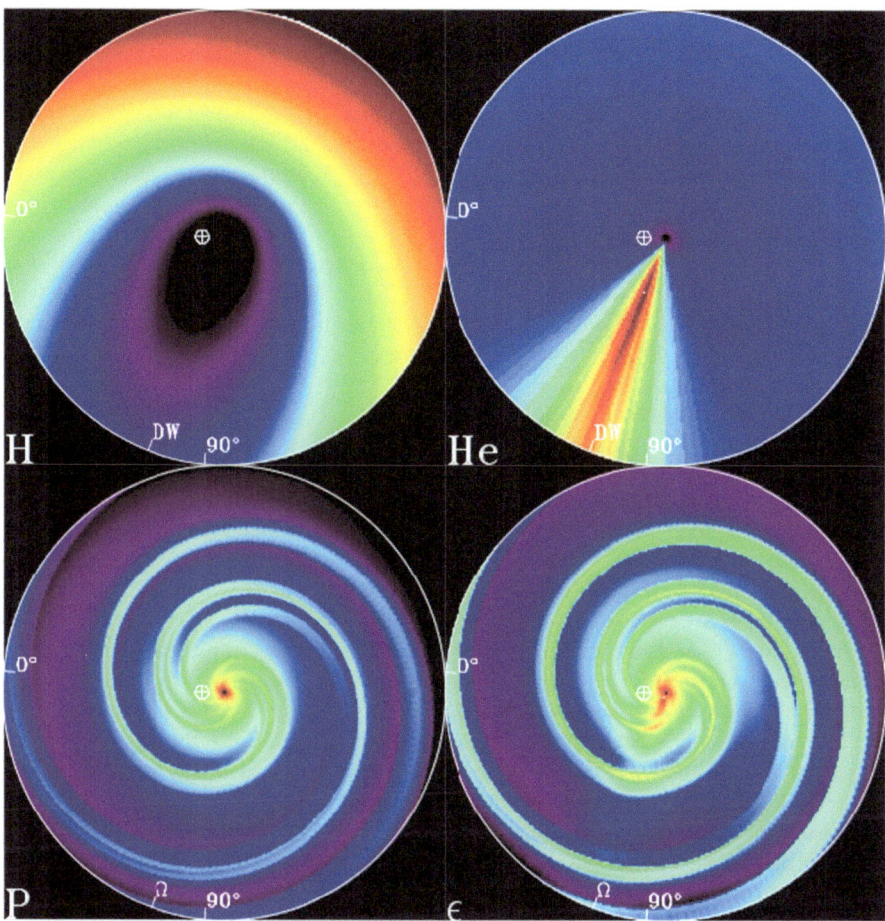

Fig. 23 The Sun-centered images showing cuts through the ecliptic plane out to a radius of 10 AU, with longitudes of 0° and 90° labeled, the location of the Earth noted, and the downwind direction marked by DW. *Upper Left*: The modeled H density distribution with values running from 0 cm^{-3} (black/purple) to 6.1×10^{-2} cm^{-3} (red/white). *Upper Right*: The modeled He density distribution with values running from 0 cm^{-3} to 5.2×10^{-2} cm^{-3}. The simulated densities are based on models by Dalaudier et al. (1984) and Lallement et al. (2004a) for He and Lallement et al. (1985a,b) for H. *Lower Left*: The solar wind proton density as modeled by ENLIL with a logarithmic color scale ranging from 10^{-2} cm^{-3} to 10^3 cm^3. *Lower Right*: The relative X-ray emissivity, $\epsilon = (n_{\mathrm{H}} + F n_{\mathrm{He}}) \, n_{swp} \, V_{rel}$, where the logarithmic scale runs from 10^4 to 10^8 cm^{-5} s^{-1} (Kuntz et al. 2015). F is a scale factor near unity that accounts for the small difference between the interaction cross sections of H and He

ionization cavity, the densities and sizes of the He focusing cone and cavity depend on solar activity (Lallement et al. 2004b). Figure 23 presents the predicted distributions of interstellar H and He atoms within the heliosphere during solar maximum.

The spatial and temporal variations of the high charge state population within the solar wind introduce further structure into soft X-ray emissions within the heliosphere. As discussed in Sect. 3.3, the largest and strongest heliospheric plasma density variations are those associated with corotating interaction regions and coronal mass ejections. Spiral shock fronts associated with corotating interaction regions provide factor of ~ 3 density enhancements that last ~ 1 day (Borovsky and Denton 2010), while the factor of 2 to 3 magnetosheath density enhancements (Guo et al. 2010) that precede CMEs last ~ 11 hours (Zhang et al. 2008).

Since solar wind charge exchange emissions are proportional to neutral population densities, the expected and observed azimuthal asymmetry in the heliospheric neutral He densities implies an asymmetry in heliospheric soft X-ray emission, a topic further explored in Sect. 5.4. The arrival of solar wind structures with different densities and compositions results in time- and spatial dependent variations in soft X-ray emissions superimposed upon those due to the neutral density asymmetries. Since the emission spectrum is comprised of many lines from 150 to 350 eV (e.g., see Fig. 15), it is difficult for detectors with low or medium spectral resolution to distinguish this spectrum from the thermal spectrum of the local bubble (see Sect. 4.4).

4.3 Planetary Emissions

Processes other than solar wind charge exchange can generate soft X-rays at Venus, Mars, Earth, and Earth's Moon. Jupiter and Saturn present particularly interesting cases. Here charge exchange produces soft X-ray aurorae, electron bremsstrahlung dominates the auroral spectra at energies above 3 keV, and the brightness of the planetary disk in soft X-rays varies proportionally to that of solar X-rays. The latter is particularly noticeable when individual solar flares are mirrored in the Jovian soft X-ray light curve.

The Einstein Observatory provided the first detection of X-rays with energies 0.2–3.0 keV from Jupiter's aurora. Metzger et al. (1983) proposed that this emission is related to energetic ion precipitation and noted that either a combination of O and S line emissions or electron bremsstrahlung continuum could fit the spectral data. Since the electrons probably cannot input sufficient power and the spectra are too soft for X-ray emission due to bremsstrahlung, line emissions from O and S heavy ion precipitation was taken to be a more likely cause of the X-ray aurora. ROSAT observations in the early 1990s confirmed this general picture (Waite et al. 1994). Horanyi et al. (1988) initially modeled the auroral X-ray emissions in terms of precipitating low ($q < 4$) charge state O and S ions. Cravens et al. (1995) subsequently invoked charge exchange with highly charged O ions.

The precipitating ions could originate in the magnetosphere (e.g., Io's volcanoes) or solar wind. It should be possible to distinguish between these sources by inspecting the charge exchange emission lines in the 0.3–0.4 keV band, where the presence of S lines would indicate an Io origin, whereas C lines would indicate a solar wind origin. Cravens et al. (2003) concluded that both cases require substantial particle acceleration to produce the observed X-ray fluxes.

Surprisingly, Chandra X-ray observatory (Weisskopf et al. 2002) observations of Jupiter in 2000 revealed that the polar hot spot of X-ray emissions pulsates with a well defined 45 minute period and maps magnetically to distances exceeding 30 R_J from the planet (Gladstone et al. 2002). Subsequent Chandra observations indicate much less organized periodicities ranging from 20 to 70 min (Elsner et al. 2005). Bunce et al. (2004) attributed these periodicities to particle acceleration driven by pulsed reconnection at the dayside magnetopause.

For "fast flow" solar wind conditions with high plasma densities and IMF strengths, and potential drops of ~ 5 MV that strip the ions (e.g., Cravens et al. 2003), precipitating magnetospheric O ions can produce X-ray intensities that match the observations. The heavy ion precipitation should be associated with fluxes of relativistic electrons escaping Jupiter (Ozak et al. 2013) as well as significant downward field-aligned currents. The current Juno mission will no doubt shed much light on magnetosphere-ionosphere interactions at Jupiter.

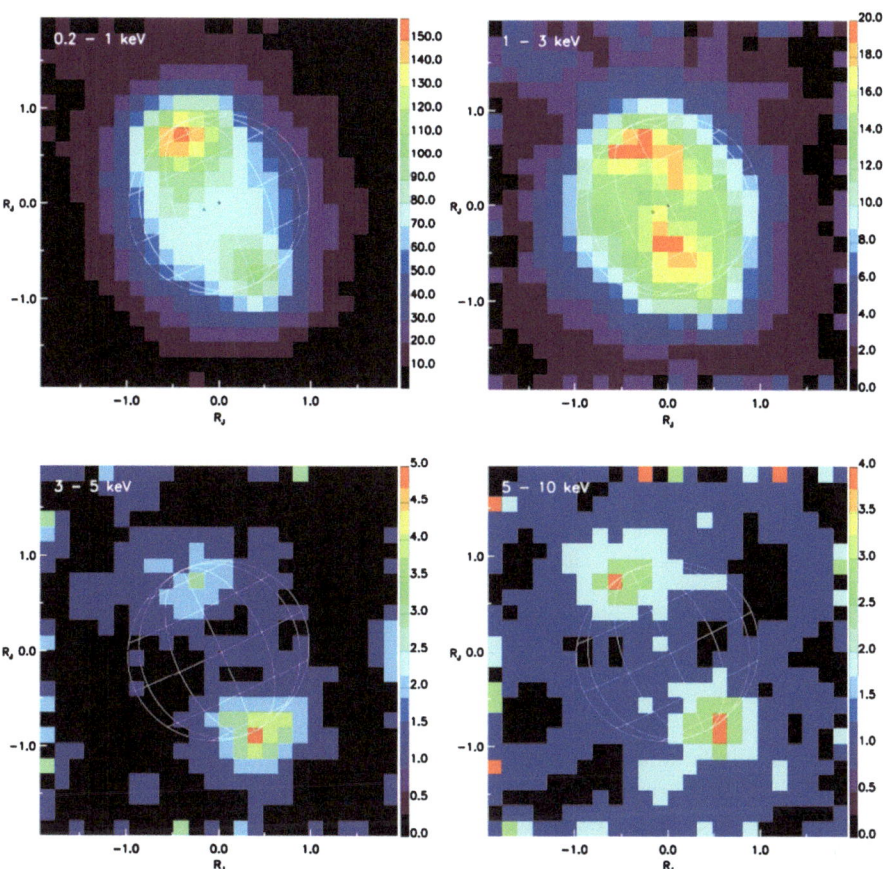

Fig. 24 XMM-Newton EPIC CCD images of Jupiter in narrow energy bands. The top left and two bottom panels show auroral contributions in the 0.55–0.60 keV O^{6+} line, at 3–5 keV, and at 5–10 keV, while the top right panel shows the disk contribution in the 0.70–0.75 keV and 0.80–0.85 keV Fe^{17+} lines (Branduardi-Raymont et al. 2007). The color scale bar is in units of EPIC counts

XMM-Newton (Jansen et al. 2001) and Chandra X-ray observations demonstrate that line emissions (in particular O^{6+}) resulting from charge exchange between highly stripped energetic ions and H_2 molecules in the planet's upper atmosphere dominate Jupiter's auroral X-ray spectrum below 2 keV (Branduardi-Raymont et al. 2004, 2007; Elsner et al. 2005). At higher energies (2–10 keV), the featureless auroral X-ray spectrum can be attributed to electron bremsstrahlung. Figure 24 shows XMM-Newton EPIC spectral maps in narrow energy bands: the aurorae are very evident in the lower energy band centered on the charge exchange O^{7+} line (top left panel) and the higher energy bands where bremsstrahlung dominates (bottom panels). By contrast, a round uniform disk is observed for a band centered on the Fe lines (top right panel) that characterize the scattered solar coronal spectrum. Unfortunately neither XMM-Newton nor Chandra possess the combination of collecting area and high resolving power needed to separate C from S lines in the 0.3–0.4 keV band. There are some indications from XMM-Newton and Chandra (Branduardi-Raymont et al. 2007; Hui et al. 2009, 2010a; Ozak et al. 2010) that S lines may provide a better spectral fit at low energies, implying ions of magnetospheric origin.

Fig. 25 Polar projection of Chandra soft charge exchange (< 2 keV, small green dots) and hard electron bremsstrahlung (> 2 keV, large green dots) X-ray emission events superimposed on a simultaneous Hubble Space Telescope Imaging Spectrograph UV image (orange) of Jupiter's aurora (Branduardi-Raymont et al. 2008). The 10° spaced grid is fixed in System III with longitude 180° toward the bottom and longitude 90° to the right. The system III z-axis lies along the planet's spin axis and longitude increases from East to West according to an observer at Earth

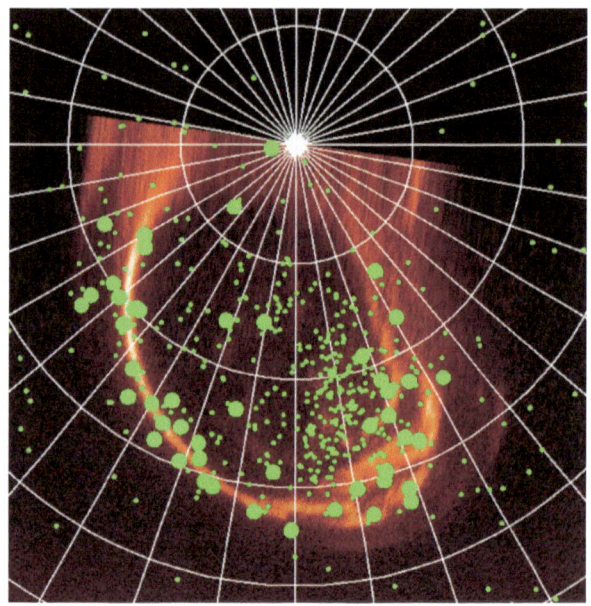

Electrons are believed to produce ultraviolet auroral features at Jupiter. As shown in Fig. 25, Chandra observations demonstrate that high (but not low) energy X-ray emissions tend to occur over the ultraviolet auroral oval and other bright ultraviolet features, suggesting that the 10–100 keV precipitating electrons that excite atmospheric H and H_2 also produce high energy X-ray emissions (Branduardi-Raymont et al. 2008). Since Jupiter's ultraviolet aurora brightened following the arrival of a solar wind shock (Clarke et al. 2009), it would not be too surprising for the electron bremsstrahlung X-ray emission to follow the same trend. And indeed, the electron bremsstrahlung spectral component varied significantly over a 3.5 day XMM-Newton observation in 2003 November following the "Halloween Storm", probably in response to magnetospheric particle energization caused by a compression of the Jovian magnetosphere corresponding to the arrival of a solar wind shock or a CME.

Chandra observed a brightening of auroral X-ray emissions, mostly below 0.5 keV, around the time when a CME was expected to impact Jupiter in October 2011. As indicated by the map shown in Fig. 26, a comparison with magnetic field models (Vogt et al. 2011) indicates that these emissions tend to cluster at the footprints of open magnetic field lines that map to the outer magnetosphere. This led Dunn et al. (2016) to associate the origin of at least some of the X-ray emissions with possible direct solar wind O and C precipitation.

Saturn, like Jupiter, emits powerful auroral emissions at radio, infrared, and ultraviolet wavelengths. Solar wind compressions cause ultraviolet and radio brightenings (e.g., Clarke et al. 2009). By analogy with Jupiter, X-ray aurorae powered by charge exchange should also be expected on Saturn, yet none have been observed to date. No auroral X-rays were observed at the time when a CME was predicted to arrive (Branduardi-Raymont et al. 2013), perhaps because the required accelerating potentials were absent (Hui et al. 2010b). Nevertheless, the planet is a source of soft X-rays. A combination of elastic and fluorescent scattering of solar X-rays in the H_2 and CH_4 atmosphere can account for Saturn's disk, polar cap and ring emissions.

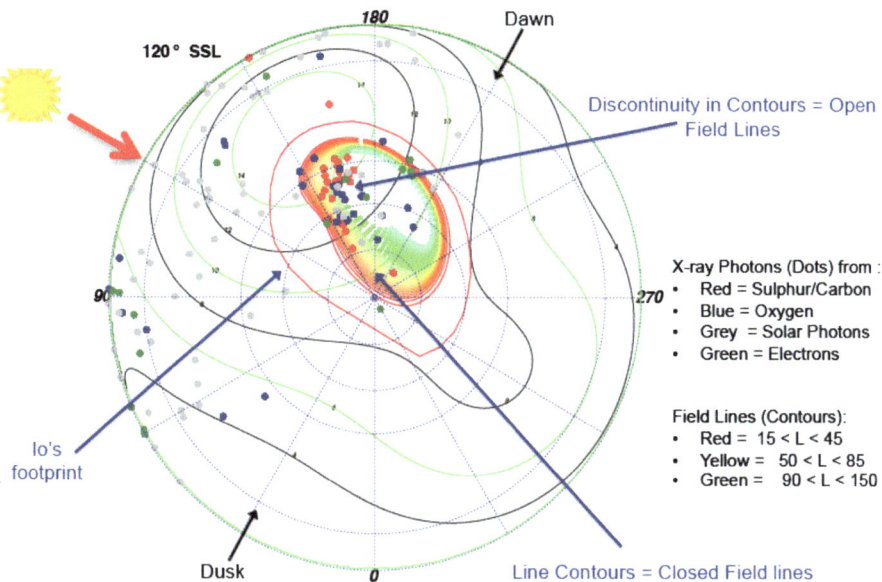

Fig. 26 Chandra X-ray events (colored dots, plotted in Jupiter's System III coordinate system: colors and species are shown on the RHS) gather at the footprints of open field lines in Vogt et al. (2011) model (colored oval). The red arrow indicates the Sun/noon position (from Dunn et al. 2016)

As for Jupiter, the flux from the disk tracks that of solar X-rays (Bhardwaj et al. 2005a,b; Branduardi-Raymont et al. 2010). Fluorescence also explains most of the disk emissions observed from Venus and Mars. Here solar X-rays ionize and remove K-shell electrons from the C and N atoms in atmospheric neutrals like CO_2 or N_2 (while on Jupiter and Saturn this occurs for the C in CH_4). The emissions occur when the K-shell vacancy is filled by a higher energy valence electron.

Looking further afield, we can expect Uranus, Neptune, and Saturn's moon Titan to emit X-rays by magnetospheric particle precipitation, and by scattering of solar X-rays. Nevertheless, Bhardwaj et al. (2007) reported that Chandra failed to detect X-rays from Uranus in August 2002. A comparison of the parameters relevant to aurora production at Uranus and Neptune with those for Jupiter leads to the conclusion that X-ray emissions from these planets are far too faint to be detected by current X-ray observatories, but may be just bright enough to be observed by Athena, the next generation X-ray observatory, when it flies around 2028 (Branduardi-Raymont et al. 2010). There is even some evidence for X-ray emission from Pluto (Lisse et al. 2017).

In addition to charge exchange, there are other non-thermal mechanisms for generating soft X-rays in solar system environments much colder than the solar corona. These environments include the Earth's upper atmosphere, comets, and the Jovian upper atmosphere where neutral temperatures are only ~ 1000, 20, and 300 K, respectively. Electron temperatures in the ionospheres of these bodies are somewhat higher than these neutral temperatures, but not by more than a few thousand degrees. Consequently, the X-ray emissions from these targets result from non-thermal processes and not thermal collisions. For example, the precipitation of electrons accelerated to high energies in the Earth's magnetotail produces bremsstrahlung X-rays in the Earth's auroral oval. In addition, planetary atmospheres scatter solar X-rays through both Thompson and fluorescent processes (e.g., Schmitt et al. 1987; Snowden and Freyberg 1993).

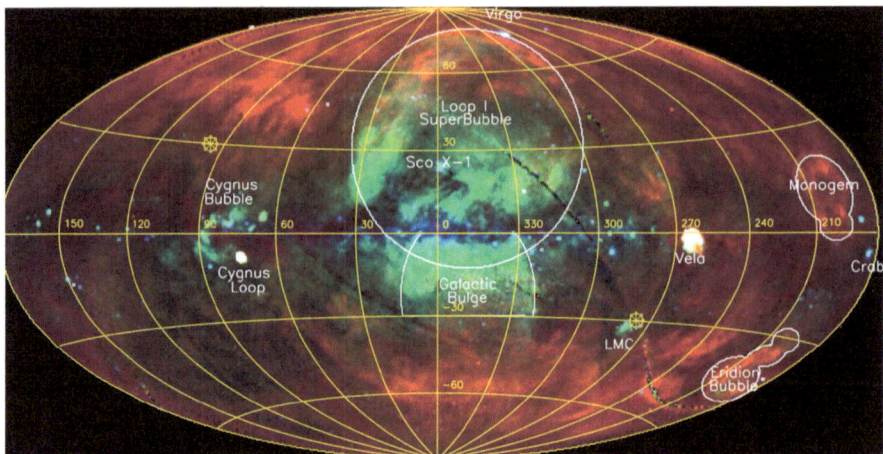

Fig. 27 False color image of the soft X-ray background with Red: $\frac{1}{4}$ keV emission, Green $\frac{3}{4}$ keV emission, and Blue 1.5 keV emission. The plot is in Galactic coordinates centered on the Galactic Center and the coordinate grid marked every 30 degrees. Stars indicate the north ecliptic pole (in the northern Galactic hemisphere) and the south ecliptic pole (in the southern Galactic hemisphere). The bulk of the X-ray point sources have been removed; the remaining point-like sources near the Galactic plane are mostly supernova remnants, while those at high latitudes are usually clusters of galaxies. Outlined or otherwise identified are the Virgo cluster of galaxies, the Loop I superbubble, the Eridanus-Orion bubble (otherwise known as the Eridion bubble), the Cygnus superbubble, the Crab, Vela, and Cygnus Loop supernova remnants, the Galactic X-ray bulge, the nearby galaxy LMC, and the Sco X-1 neutron star (the brightest X-ray source in the sky as seen from Earth). The Galactic halo emission is the red emission at Galactic latitudes above 30°

4.4 Cosmic Soft X-Ray Emissions

The cosmic X-ray sky comprises numerous components that are strongly spectrally and spatially variable. The emissions from some of these components are comparable to or brighter than those typically generated by the solar wind charge exchange processes that occur much closer to Earth. To conduct the science outlined within this paper, these background emissions must be quantified and then subtracted from soft X-ray images. Here, we consider diffuse, point, and distinct extended sources (see Fig. 27).

First observed in the late 1960s (e.g, Bowyer et al. 1968), non-heliospheric diffuse emissions originate at locations ranging from the local interstellar medium to cosmological distances (e.g., McCammon and Sanders 1990). Galactic emissions from a thermal plasma within the Local Hot Bubble (e.g., Tanaka and Bleeker 1977; Sanders et al. 1977; Snowden et al. 1990, 2014; Snowden 2002), a low density (0.05 cm^{-3}), high temperature (10^6 K), region extending from ~ 30 to ~ 150 parsecs from the Sun depending on direction, dominate the flux of soft X-rays at lower ($\sim \frac{1}{4}$ keV) energies. High neutral H column densities within the Milky Way disk absorb emissions from greater distances. There are contributions from the lower Galactic halo at high Galactic latitudes (e.g., Kuntz and Snowden 2000). Consequently, as shown in the upper panel of Fig. 28 (Snowden et al. 1997), the surface brightness at $\frac{1}{4}$ keV generally increases from the Galactic equatorial plane towards both the north and south poles. While some of the features superimposed upon this general pattern can be identified with specific Galactic objects, most result from integral LOS filling factor and density (i.e., emission-measure) variations in diffuse $\sim 10^6$ K plasmas further strongly modified by variable absorption in the interstellar medium.

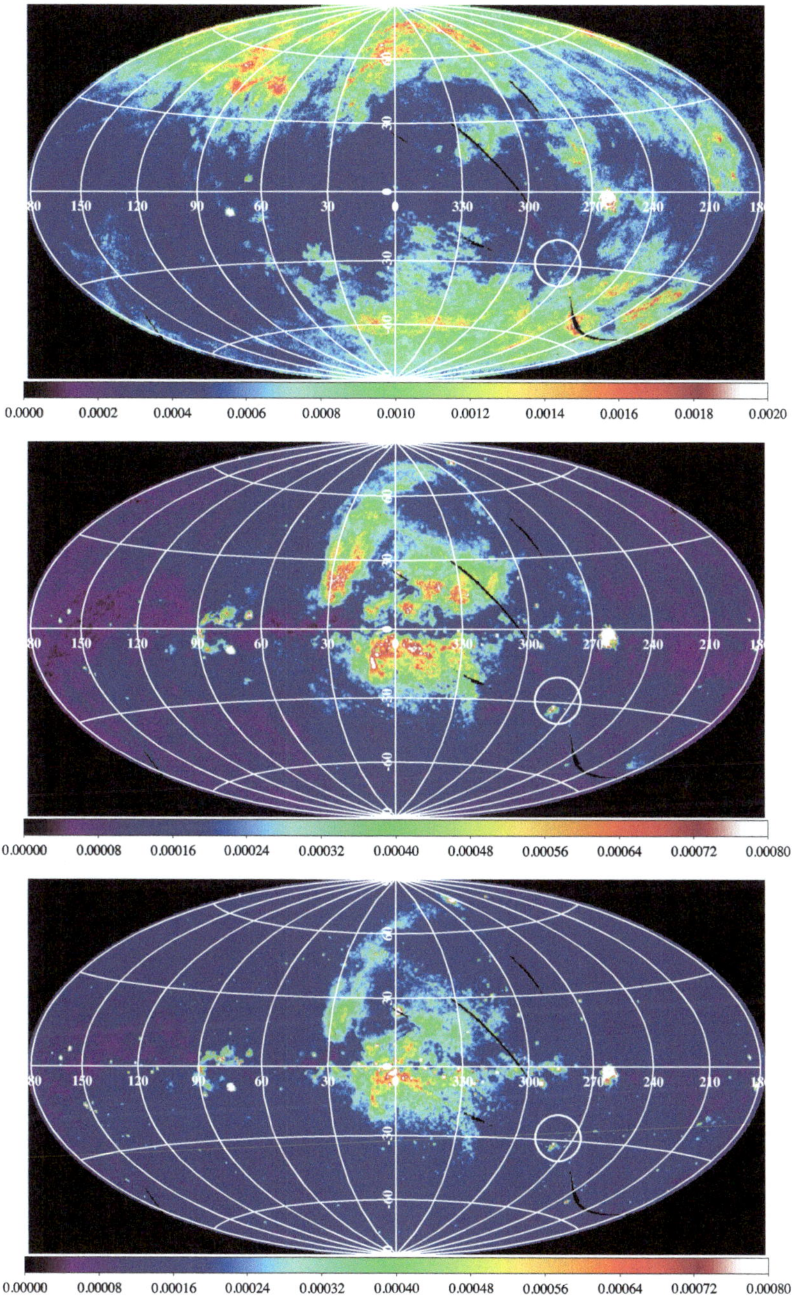

Fig. 28 ROSAT All-Sky Survey (RASS, Snowden et al. 1997) images of the soft X-ray background shown as Aitoff-Hammer equal-area maps in Galactic coordinates centered on the Galactic center. Galactic longitude increases to the left, the south Galactic pole is at the bottom and the north Galactic pole is at the top. Purple and blue indicate low intensity while red and white indicate high intensity. The units of the color bars are ROSAT counts s^{-1} arcmin^{-2}. *Upper Panel*: $\frac{1}{4}$ keV band; *Middle Panel*: $\frac{3}{4}$ keV band; *Lower Panel*: 1.5 keV band. The white circle in the lower right of all figures outlines a 10° radius region surrounding the south ecliptic pole

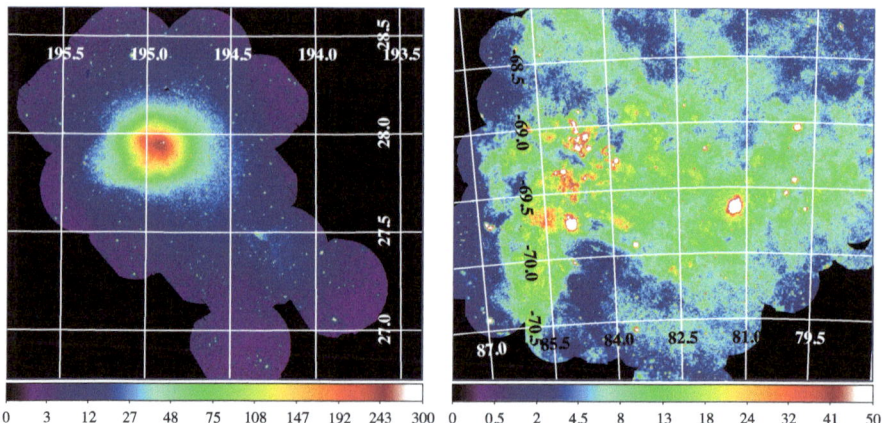

Fig. 29 *Left Panel*: XMM-Newton mosaic of the Coma Cluster of galaxies in the 0.4–1.25 keV band. The data are square root scaled and the units are in counts s^{-1} deg^{-2}. The coordinates are in right ascension and declination. *Right Panel*: XMM-Newton mosaic of a region of the Large Magellanic Cloud, a satellite galaxy of the Milky Way, in the same band, coordinates, and units (unpublished images provided by S.L. Snowden)

At the higher energies ($\frac{3}{4}$ keV) shown in the middle panel of Fig. 28, distinct objects (some extending over large solid angles) dominate a relatively flat background. The strong enhancement at the Galactic center represents the combined emission from the nearby Loop I Superbubble and the Galactic Bulge (e.g., Snowden et al. 1997). The generally flat background combines the extragalactic background (primarily unresolved point sources) with emissions from the Galactic halo and our local group of galaxies modulated by absorption from the interstellar medium near the Galactic plane. The 1.5 keV band map is shown in the lower panel of Fig. 28. It shows a structure similar to the $\frac{3}{4}$ keV band map except that the distinct features are not as bright.

As can be seen in the maps, the cosmic diffuse X-ray background (e.g., Snowden et al. 1997) is bright and spatially varying, differs radically in the $\frac{1}{4}$ keV and $\frac{3}{4}$ keV bands, and is observed in all directions. It will therefore be present in all observations of soft X-rays generated by solar wind charge exchange. However, it is temporally constant on all human time scales and is well understood and mapped. Consequently it can be subtracted from the light curves and images of soft X-rays generated by charge exchange observations. While the surface brightness of the cosmic background varies by up to an order of magnitude when small regions are considered, the south ecliptic pole (a likely background direction for a soft X-ray mission imaging the subsolar bow shock and magnetopause from a polar vantage point) lies in a relatively benign direction, particularly at $\frac{1}{4}$ keV. At $\frac{3}{4}$ keV, the Large Magellanic Cloud (a nearby galaxy seen in the right panel of Fig. 29) does show enhanced emissions but it has been particularly well studied in soft X-rays (e.g., Snowden and Petre 1994; Haberl 2014).

In addition to the diffuse emission, there are many point sources, including stars, compact objects (e.g., pulsars, X-ray binaries), and active galactic nuclei (AGN). Figure 30 shows the locations of the 18,811 bright sources (Bright Source Catalog, BSC) detected during the ROSAT All-Sky Survey (RASS) (Voges et al. 1999). A further 105,924 sources were included in the Faint Source Catalog extension to the BSC. Much of what appears to be a general diffuse emission at higher energies in Fig. 28 (e.g., at high latitudes in the middle and lower panels) actually results from the superposition of unresolved emission from AGN at cosmological distances. These individual sources are insufficiently bright to be detected

Fig. 30 Locations, relative fluxes (the size of the circle scales as the log of the flux), and hardnesses (purple indicates a hard source, red a soft source) of the 18,811 sources listed in the RASS Bright Source Catalog (Voges et al. 1999). (Courtesy of the Max Planck Institute for Extraterrestrial Physis, Garching, Germany)

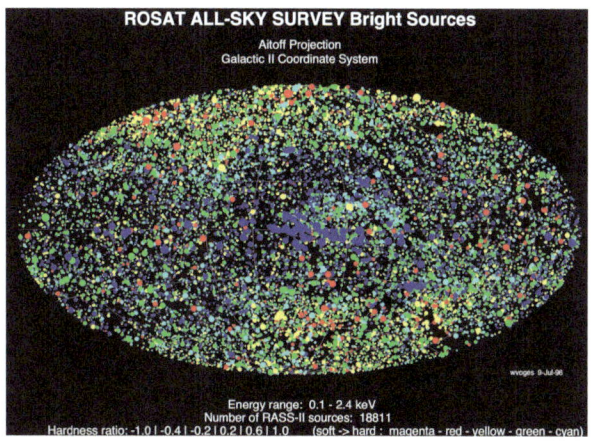

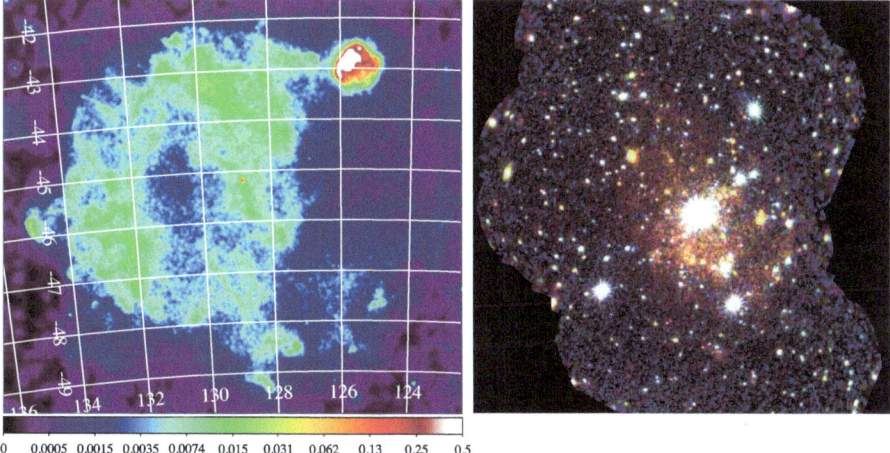

Fig. 31 *Left Panel*: RASS image of the Vela supernova remnant (SNR) along with the Puppis SNR (small bright region in the upper right of the image) in the $\frac{3}{4}$ keV band. The data are logarithmically scaled and the units are counts s^{-1} deg^{-2}. The coordinates are in right ascension and declination. *Right Panel*: Color-coded XMM-Newton image of the galaxy M33 centered at right ascension, declination = 23.46°, 30.66°. The image shows a large number of point sources along with some diffuse emission. Orange/red indicates softer sources while blue indicates harder sources with brightness indicating the relative strength of the source. The image is ∼ 0.91° from top to bottom (unpublished images provided by S.L. Snowden)

with the available exposure time allowed by the RASS, but can be resolved by newer observatories particularly during deep observations like the XMM-Newton image of the Hubble Deep Field North shown in the right panel of Fig. 31. As shown in the upper panel of Fig. 28, distinct galactic supernova remnants (SNRs) such as Vela and Puppis (enlarged in the left panel of Fig. 31) can subtend relatively large areas on the sky and can be both bright and strongly spatially varying. Nearby galaxies and clusters of galaxies can contribute as both point and extended sources, sometimes spectacularly as illustrated by the examples in Fig. 29.

Because they do not vary with time and are reasonably well mapped, extended sources, even small-scale ones, can be modeled and subtracted from observations of solar wind

Table 5 Soft X-ray emissions at Earth

Region	LOS dimension L (R_E)	Neutral density n_N (cm^{-3})	Ion density n_{SW} (cm^{-3})	High charge state ion fraction (%)	Effective velocity[a] V_{eff} (km s^{-1})	Relative LOS emission strength
Foreshock	10	10	5	0.1–0.2	400	30–60
Dayside magnetosheath	5–10	20–40	15–20	0.1–0.2	400	90–1000
Flank magnetosheath	10	6–12	10–12	0.1–0.2	400	45–180
LLBL/PDL	1–2	20–40	5	0.1–0.2	400	6–50
Cusps	1	20–400	5–10	0.1–0.2	400	6–500
Ring current radiation belts	5	100–200	1	10^{-4}–10^{-2}	4000	0–60
Plasmasphere	6–8	200–400	10–10^3	0	40	0

[a]$V_{eff} = (V_{th}^2 + V_{bulk}^2)^{\frac{1}{2}}$

charge exchange emissions. On the other hand, point sources can vary greatly with time. The difficulty of removing point sources from SWCX data depends to a certain degree on the characteristics of the observing instrument, such as the point-spread function of its optics and to a lesser extent on the energy resolution of its detectors. Fortunately, nearly all point sources are relatively dim when compared to the cosmic diffuse background for any angular binning greater than a few arc minutes, and can typically be included in the cosmic background. The brighter sources have been well documented (e.g., by the RASS BSCs), and as long as the point spread function of the telescope is well understood their contributions can also be modeled and subtracted, both spatially and temporally.

5 Simulations

Simulations are needed to (1) predict integrated LOS soft X-ray intensities for comparison with observations from previously-flown narrow FOV astrophysical telescopes, (2) determine how these intensities vary with changing solar wind conditions, (3) identify regions that are strong soft X-ray emitters and therefore appropriate targets for wide FOV imagers, (4) determine whether the sharp features (e.g., bow shock, magnetopause, cusp edges) that bound these targets can be identified in integrated wide FOV LOS images, (5) predict the best vantage points for future missions, and (6) estimate whether planned wide FOV telescopes will have the spatial and temporal resolution needed to track the relevant phenomena. Furthermore, we can combine simulation results with observed LOS intensities to estimate the poorly known $\frac{1}{4}$ keV band integrated cross-section for soft X-rays generated by charge exchange. Once this band-integrated cross-section is known, the intensities expected for a band-integrating detector looking in any direction can be calculated.

5.1 Relative Emission Strengths

From Eq. (6), the relative strengths of emissions from terrestrial, planetary, and interplanetary targets can be expressed as the product of the neutral density, plasma density, and effective interaction velocity integrated over the path length. Table 5 summarizes informa-

Table 6 Soft X-ray emissions at Venus, Earth, and Mars

	Venus	Earth	Mars
A. Characteristic obstacle dimension (R_{MP} or R_{planet})	6050 km	60000 km	3390 km
B. SW density	12.7 cm^{-3}	6.2 cm^{-3}	3.3 cm^{-3}
C. Exospheric density	1000 cm^{-3}	30 cm^{-3}	10000 cm^{-3}
D. $V_{effective}$	400 km s^{-1}	400 km s^{-1}	400 km s^{-1}
A*B*C*D normalized	6.9	1.0	10.0

Solar wind densities obtained from Köhnlein (1996). Exospheric H densities in the magnetosheath regions of interest were determined from Lyα observations (Futaana et al. 2011; Zoennchen et al. 2011). V_{eff} is defined as $(V_{th}^2 + V_{bulk}^2)^{\frac{1}{2}}$, where V_{th} is the thermal velocity and V_{bulk} is the bulk velocity (Robertson and Cravens 2003b). Note that V_{eff} in the magnetosheath is approximately V_{SW}, the solar wind velocity (Spreiter et al. 1966)

tion concerning ion and neutral densities, the fraction of ions with high charge states, the effective velocity of the ions, and estimates for the characteristic line-of sight dimensions of various magnetospheric regions. The final column indicates the relative intensity of the soft X-ray emissions to be expected from each region of space: this column is simply the normalized product of those in the preceding 5 columns. The calculation demonstrates that the main targets for a soft X-ray imager must be the magnetosheath and cusps. Due to the sharp differences in intensities expected along lines-of-sight that do and do not pass through the dayside magnetosheath and cusps, such an imager should be particularly effective in identifying the locations of the dayside bow shock, magnetopause, and cusps. Table 6 compares expected soft X-ray intensities from Venus and Mars with those from Earth. Greater exospheric neutral densities in the solar wind interaction regions make the magnetosheaths of Venus and Mars far brighter targets than that of Earth.

Table 7 compares measured soft X-ray emission intensities from the soft X-ray diffuse background, comets, the Moon, Mars, Venus, and Jupiter.

5.2 Past Modeling Efforts

Robertson and Cravens (2003b), Robertson et al. (2005, 2006, 2012, 2013) reported a series of papers modeling the soft X-rays emitted from the Earth's magnetosheath and cusps. They employed Eq. (6), either the Spreiter et al. (1966) gasdynamic or the BATS-R-US magneto-hydrodynamic (MHD) (Tóth et al. 2012) models for magnetosheath and cusp densities, and the Hodges (1994) model for exospheric neutral densities to integrate line of sight intensities within the FOV of a notional instrument on a high apogee orbit. Their results demonstrated that it would be easy to identify and track the time-varying locations of cusp, magnetopause, and bow shock boundaries in the images taken by moving wide FOV soft X-ray telescopes. Although they concluded that integrated line-of-sight charge exchange soft X-ray intensities are less than those of the soft X-ray background during quiet times, they can rise to values far greater than those of the soft X-ray background during intervals of enhanced solar wind plasma fluxes.

Sun et al. (2015) employed the 3-D PPMLR global MHD plasma model (Hu et al. 2007) and a simple exospheric neutral model to define the soft X-ray signatures that would be observed by an outward-looking low-altitude spacecraft when Kelvin-Helmholtz waves move antisunward on the flanks of the magnetosphere. They showed how the amplitude, speed, distribution, and temporal evolution of the vortices can be derived from the simulated soft

Table 7 Measured soft X-ray intensities

Source	Intensity ($\mathrm{keV\,cm^{-2}\,s^{-1}\,sr^{-1}}$)	Notes
Diffuse soft X-ray background	10–30	Observed during STORM suborbital flight. Value from Krasnopolsky et al. (2004)
Earth	~2.8	Based on values reported for ROSAT observations through the flank magnetosheath (Kuntz et al. 2015). Values for the subsolar magnetosheath will be a factor of ~4 times greater
Comets	~10,000	Highly variable with heliocentric distance and cometary gas production rate and solar wind conditions (e.g., fast polar *vs.* slow equatorial, or CME Bodewits et al. 2007)
Moon	11	From ROSAT observation (Collier et al. 2014)
Mars and Venus	~3000	Assumes 1 MW halo luminosity—Observational determinations vary from 0.3–10 MW (Dennerl 2002)
Jupiter	~400,000	Jovian auroral emission attributable to charge exchange with accelerated Iogenic ions (Cravens et al. 1995; Elsner et al. 2005)

All solar system soft X-ray sources listed above have sufficient intensity to be observable by a 3U Wide FOV CubeSat Imager

X-ray signatures. Kuntz et al. (2015) used the BATS-R-US plasma and Hodges' exospheric model to simulate soft X-ray emissions from the magnetosheath and cusps (e.g., Fig. 32). Emission intensities increase faster than linearly as the solar wind flux increases because neutral densities increase rapidly as the emitting region moves deeper into the exosphere. They also showed that peak emissions along lines-of-sight that pass through the subsolar magnetosheath exceed background levels for all but the lowest solar wind plasma fluxes.

Walsh et al. (2016b) described the process by which realistic global images of soft X ray images from the magnetosheath and cusps can be simulated for wide field-of-view imagers at specified locations. First the magnetospheric contribution is calculated using plasma values from a global magnetohydrodynamic simulation and neutral densities from an exospheric model. Next components from the galactic background and the Earth are added. Finally, Poisson noise is added. Since the backgrounds are well known, they can be subtracted to produce images resulting solely from charge exchange in the vicinity of Earth. Whittaker et al. (2016) followed this procedure to simulate charge exchange emissions for comparison with XMM-Newton observations along lines-of-sight through the magnetosheath.

Section 7 provides further information concerning the capabilities of currently available instrumentation to resolve features in soft X-ray images and discusses methods to identify the boundaries with cadences appropriate to the intended magnetospheric research problems.

5.3 Modeling Soft X-Ray Emissions from Venus and Mars

Soft X-ray emission resulting from charge exchange at Venus and Mars has also been modeled. Holmström et al. (2001) employed an empirical flow model to predict emissions from Mars, showing that they should map out the dayside magnetosheath. Gunell et al. (2004) then employed a hybrid code model to simulate emissions generated by high charge state solar wind ions when they encounter the Martian exosphere. They concluded that soft X-rays

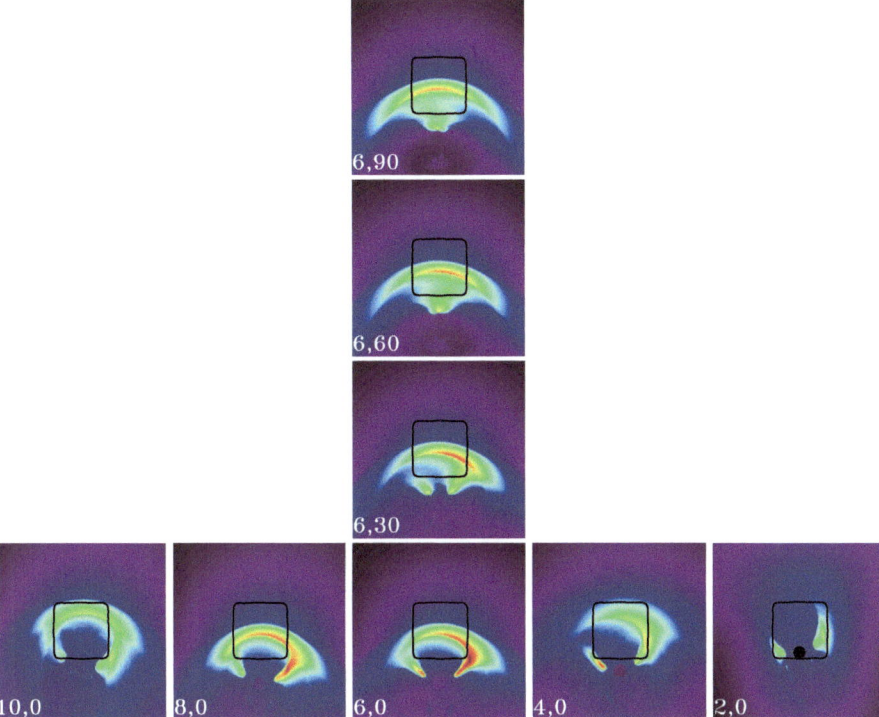

Fig. 32 Simulations of soft X-ray emissions from the magnetosheath as observed from various locations. The numbers in the lower left of each panel are the hour and ecliptic latitude of the spacecraft. In all cases the spacecraft targets GSE = [10, 0, 0]. The FOV of the baseline instrument is marked as black. Note that from a wide range of locations it is still possible to image the magnetopause, although it can get a bit faint when the observation point is too far behind the terminator

make only a small contribution to the intensity observed when viewing the disk, which is dominated by the fluorescence and scattering of solar X-rays (Cravens and Maurellis 2001). Just outside the disk, which has a radius of 10 arc seconds when viewed from Earth, soft X-rays generated by charge exchange dominate but their intensities diminish rapidly versus radial distance through a halo with a width of 20 arc seconds surrounding the planetary disk. Gunell et al. (2005) demonstrated that both the intensity of the X-ray emissions and the size of the X-ray halo increase with increasing exobase neutral temperature. Holmström (2006) concluded that the latitudinal and seasonal effects of asymmetries in exobase densities propagate outward to large distances from Mars, and can therefore be detected in soft X-ray emissions. Employing a global hybrid code simulation, Koutroumpa et al. (2012) reported that simulations predict a northern/southern hemisphere asymmetry in the intensities and locations of soft X-rays emitted from the magnetosheath. They attributed this asymmetry to differences in plasma densities resulting from motional electric fields in the solar wind, not crustal anomalies.

Finally, Gunell et al. (2007) considered the case of Venus. The greater mass of Venus than Mars causes exospheric densities to fall faster with altitude at Venus than Mars. Consequently, the halo of soft X-rays emitted by charge exchange surrounding Venus has a narrower radial extent (and is therefore more difficult to observe) than that at Mars, as shown in Fig. 33.

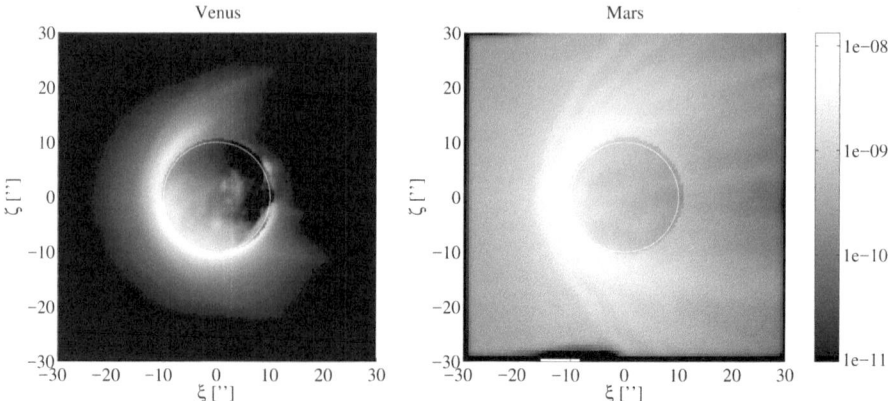

Fig. 33 Simulated X-ray images of Venus (left panel) and Mars (right panel). The vantage points are chosen such that the radius of each planet extends 10 arc seconds. The Sun lies to the left of each image. The gray scale shows the soft X-ray radiance in $\mathrm{W\,m^{-2}\,sr^{-1}}$. The white circles denote the size of the respective planets and the coordinates are shown as arc seconds as seen by an observer at Earth (Gunell et al. 2007)

5.4 Modeling Soft X-Ray Emissions from the Heliosphere

Cravens (2000a) estimated the integrated line-of-sight intensities of soft X-rays emitted by charge exchange with interstellar neutrals within the heliosphere and found them to be comparable to those from the 100 parsec wide Local Bubble, the cavity surrounding the heliosphere that is filled with a hot (10^6 K) tenuous plasma. Cravens et al. (2001) then employed a simple model for radial solar wind propagation without solar rotation to model time-dependent variations in heliospheric and geocoronal charge exchange emission. They showed that emissions from the geocorona dominate on time scales ranging from a few hours to a few days, since emissions from the small region surrounding the Earth react very quickly to solar wind stimuli. Emissions from interactions with interstellar He within the heliosphere vary on a several day time scale, while emissions from interactions with the interstellar H component contribute to a quiescent zero-level offset due to the large integrating path through the heliosphere that smooths out any variations.

Robertson and Cravens (2003a) mapped soft X-ray emissions as a function of look direction and time of year, finding relatively high emissions along lines of sight passing through the He focusing cone. Figure 34 shows that emissions increase from solar maximum to solar minimum. Koutroumpa et al. (2006) predicted that near-Earth observers should observe increases in low- and medium-latitude emissions during solar minimum, particular in December, and that these emissions should peak closer to the Sun as the sizes of the neutral cavities surrounding the Sun diminish. Furthermore, they showed that the contributions of the heliosheath and heliotail to the integrated line of sight heliospheric emissions never exceeds 5%. Parallax effects and the specific observation geometry precluded all but a fraction of the emissions from the He focusing cone from being observed by the RASS (Lallement 2004; Koutroumpa et al. 2009a, Fig. 10). Nevertheless, recent results from the December 2012 Diffuse X-rays from the Local galaxy (DXL) rocket flight have conclusively demonstrated that $\sim 40\%$ of background emissions in a direction of very low cosmic background intensity originate from the He focusing cone and not the local hot bubble surrounding the heliosphere (Galeazzi et al. 2014; Snowden et al. 2014, 2015; Uprety et al. 2016; Liu et al. 2017).

Fig. 34 X-ray emissivity for O^{7+} interacting with neutral He within the heliospheric focusing cone for (upper panel) solar minimum and (lower panel) solar maximum (similar to Koutroumpa 2012). The black circle indicates Earth's orbit. The emissivity from the He focusing cone trails the Sun at the center of the figure, which is moving towards the left. Note the differing scales for the two panels: emissivities are expected to peak during solar minimum. The view is from the North Ecliptic Pole

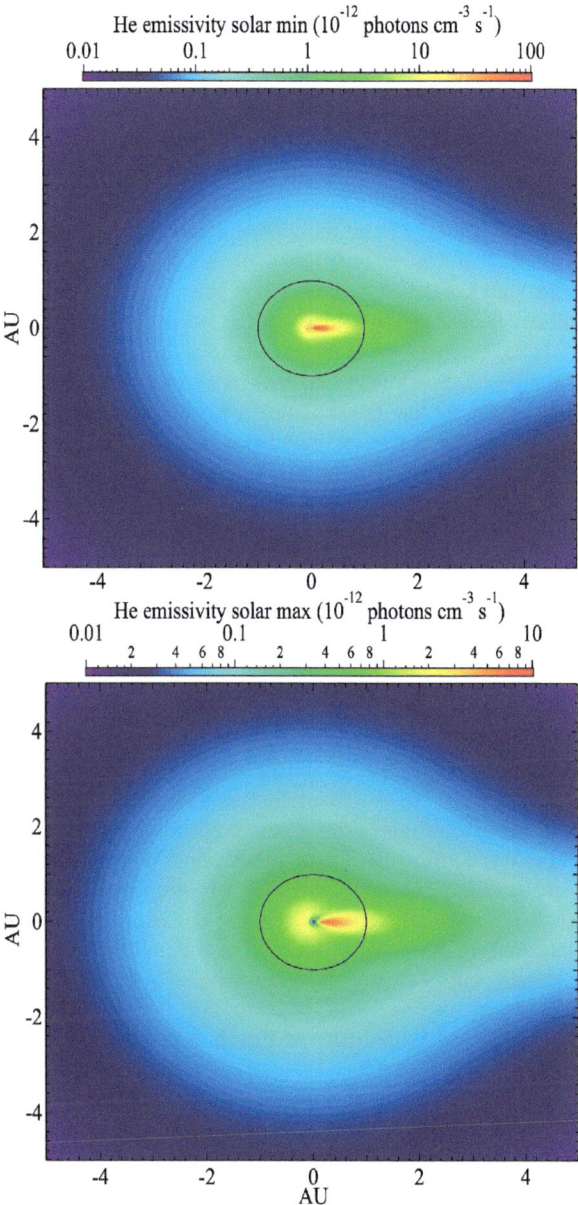

Koutroumpa et al. (2007) employed a dynamic model to simulate the effects of a prop-agating corotating interaction region on line of sight soft X-ray emissions. The CIR was simulated by a step function change in ion fluxes along a Parker spiral. Koutroumpa (2012) employed this model and the fluxes of high charge state ions observed by the Advanced Composition Explorer (ACE) to predict realistic solar wind charge exchange soft X-ray light curves which agree well with XMM-Newton, Suzaku X-ray observatory (Koyama et al. 2007) and Chandra observations in the O lines. Kuntz et al. (2015) employed more sophisti-cated models for heliospheric plasma (Odstrcil 2003) and neutral (Koutroumpa et al. 2006)

densities to describe the soft X-ray emissions seen as a function of look direction during the passage of solar wind shock fronts.

The next step for such simulations is to improve the magnetohydrodynamic models for the propagation of solar wind features and include information concerning the populations of high charge state solar wind ions. As noted in Sect. 3.3, the composition and charge states of the solar wind ion population vary with time. In particular, the ion composition of fast and slow solar wind flows are markedly different. For the fast solar wind, low charge state ions such as C^{5+} and O^{6+} are more prominent (Schwadron and Cravens 2000). For the slow solar wind, high charge state ions such as C^{6+} and O^{7+} become more prominent. Obtaining a more complete database of charge exchange collision cross-sections and line emission probabilities is needed, particularly in the lower energy range (0.1–0.3 keV) where calculations are based on hydrogenic approximations due to the complexity of the electronic structures of the Fe, Si, S, and Mg ions involved in soft X-ray charge exchange collisions. In principle, the spectral line ratios of soft X-ray emission generated by charge exchange could then be used to remotely sense abundance and velocity variations throughout the heliosphere, as already demonstrated in the case of cometary emissions (Beiersdorfer et al. 2001).

6 Past Observations of X-Rays Generated by Charge Exchange Processes

From the discussion above, it is expected that charge exchange soft X-ray emissions are generated in the vicinity of the Earth, comets, Venus, and Mars at locations where solar wind ions encounter exospheric neutrals. In addition, it is expected that there will be emission throughout interplanetary space where the targets are neutrals entering the heliosphere from the interstellar medium. At Earth, the most intense emissions should therefore occur in the dayside magnetosheath and cusps. The time-varying intensity of each line emission should depend upon the time-varying flux of corresponding high-charge state solar wind ions.

This section summarizes observations of soft (0.1–2.0 keV) X-rays by astrophysical telescopes that confirm all of these predictions. When present, soft X-ray emissions from charge exchange fill the entire narrow FOV of astrophysical telescopes with a constant foreground. Much of the work on charge exchange emissions generated in the vicinity of Earth has been motivated by the need to remove these emissions from observations of nearby supernova remnants, galaxy clusters, and the cosmic X-ray background.

6.1 Temporal Variability at Earth

ROSAT undertook an all-sky survey (Snowden et al. 1997; Voges et al. 1999) of the diffuse soft X-ray background covering the 0.1–2.0 keV band, with the energy range typically separated into three broad energy bands with mean energies of roughly $\frac{1}{4}$, $\frac{3}{4}$, and 1.5 keV. Launched on 1990 June 1, the mission carried an X-Ray Telescope with a $\sim 2°$ FOV, large effective area, and short focal length. In conjunction with the low-background Position Sensitive Proportional Counter detector, the mission achieved an unprecedented signal-to-noise ratio for diffuse emission studies. From its low-altitude (initially 580 km) nearly circular orbit with an inclination of 53°, the XRT observed soft X-ray emissions in a plane roughly perpendicular to the Sun-Earth line passing through the ecliptic poles, i.e., along lines of sight that pass through the terminator magnetosheath.

As illustrated in the upper panel of Fig. 35, curious streaks in emission are occasionally superimposed upon the diffuse soft X-ray sky seen in all-sky survey maps (Snowden et al. 1995). These streaks, called Long Term Enhancements or LTEs (Snowden et al. 1994), last from hours to days with intensities that occasionally reached or exceeded those of the cosmic

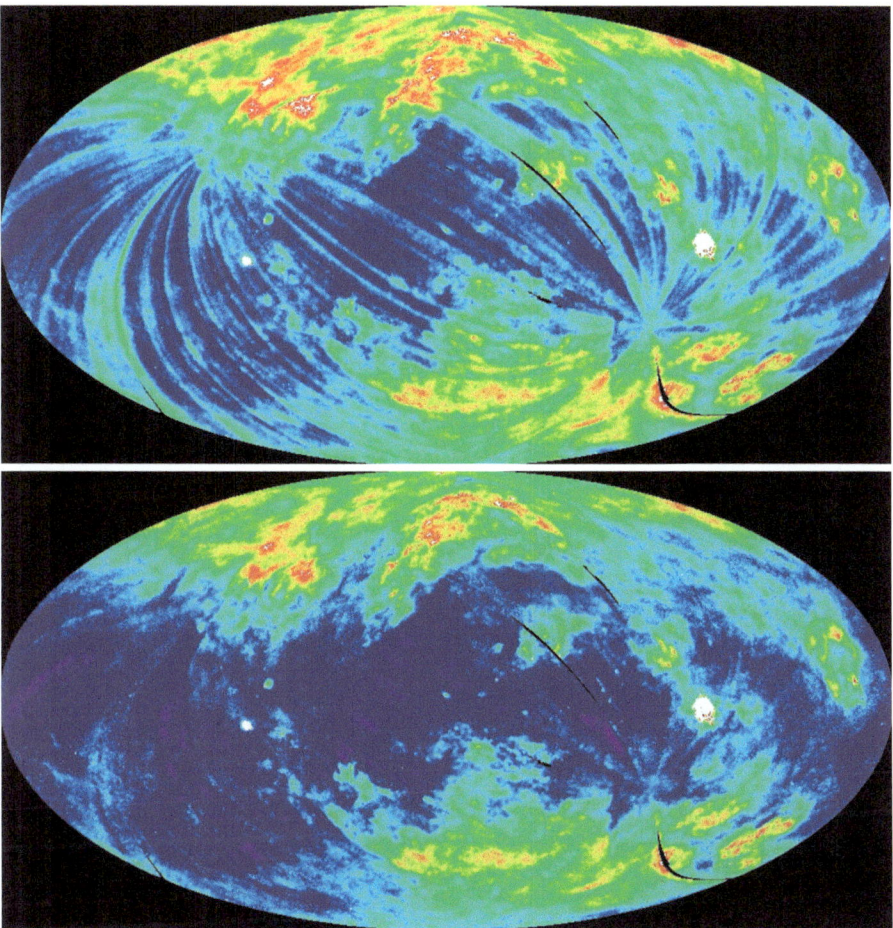

Fig. 35 *Upper Panel*: RASS map of the $\frac{1}{4}$ keV sky before the removal of Long Term Enhancements (LTEs). The map is in Galactic coordinates with $(l, b) = (0°, 0°)$ at the center and longitude increasing to the left. The diffuse X-ray surface brightness increases from purple to white (the counts rate varies from purple, ~ 250 counts s^{-1} arcmin^{-2}, to white, ~ 2500 counts s^{-1} arcmin^{-2}). *Lower Panel*: The same data after the subtraction of an empirical estimate for the LTE contribution

background. They were found to be strongest, and most common, in the $\frac{1}{4}$ keV band. By comparison with the nearly identical areas of the sky observed during adjacent orbits, the LTEs were removed from the observations to construct the all-sky survey shown in Fig. 35 (lower panel).

Cravens et al. (2001) showed that the streaks occurred during intervals of enhanced solar wind ion flux. Figure 36 compares the ROSAT $\frac{1}{4}$ keV soft X-ray light-curve from 1990 to 1991 (black) with the solar wind ion flux ($n_{sw}v_{sw}$) simultaneously observed by IMP-8 (Kuntz et al. 2015). Data points are averaged to 95 min, the approximate period of the ROSAT orbit. Short gaps in the light-curve result primarily from either non-survey operations or satellite operational issues. The long gaps in the ROSAT light curve are due to the survey being accomplished in four separate time intervals. Gaps in the IMP-8 observations reflect the absence of telemetry and times when IMP-8 was within the magnetotail or magnetosheath.

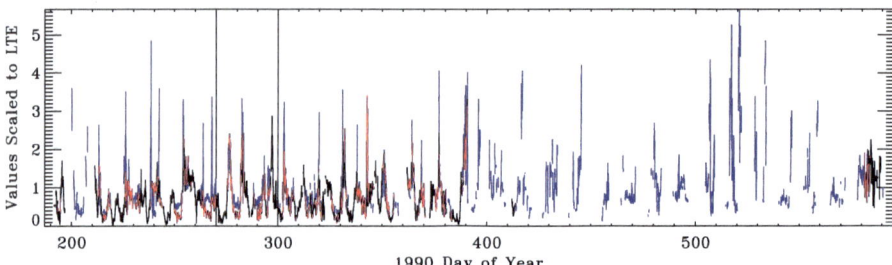

Fig. 36 $\frac{1}{4}$ keV LTE flux (black) and the solar wind flux (blue) as a function of date (Kuntz et al. 2015). The LTE flux is in units of ROSAT counts s^{-1} FOV^{-1}. The solar wind flux from IMP-8 has been scaled to match the LTE flux. Periods of LTE data for which there is also solar wind data are plotted in red. The vertical lines bound the much shorter interval considered by Cravens et al. (2001)

Fig. 37 ROSAT observations of the $\frac{1}{4}$ keV LTE flux versus IMP-8 observations of the local solar wind flux. The solid line is the best fit of (LTE count rate versus $n_{sw}v_{sw}$), the dashed line is the best fit minimizing the distance orthogonal to the fitted line, and the dotted lines show the relations with 0.54 and 1.46 times the best fit slope (Kuntz et al. 2015)

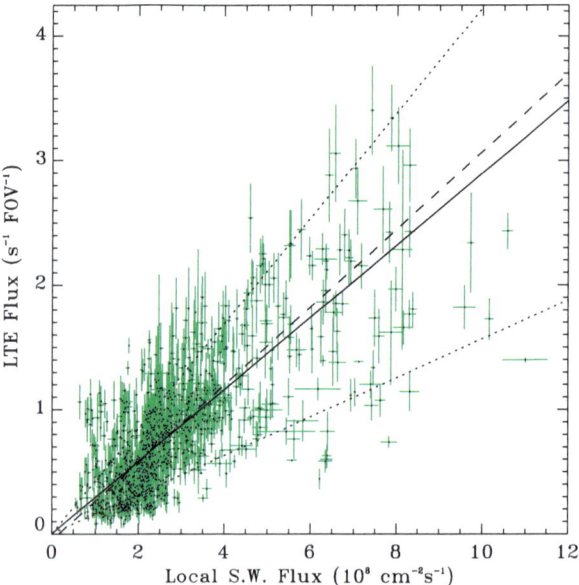

Inspection of Fig. 36 reveals that band-integrated emissions in the $\frac{1}{4}$ keV band track solar wind fluxes closely despite the variations in solar wind composition that occur as a function of solar wind structure. The scatter plot of LTE emission versus solar wind flux shown in Fig. 37 confirms this point (Kuntz et al. 2015). Here the dotted lines bounding the distribution show the range of variations expected for identical solar wind conditions but different look directions along each 95 min ROSAT orbit. The light-curve and scatter plots for the $\frac{3}{4}$ keV band (not shown) do not provide any evidence for a similar correlation with solar wind fluxes, suggesting that the O line emissions that dominate the $\frac{3}{4}$ keV band are less well correlated with the solar wind flux than the aggregate of the many lines that produce the $\frac{1}{4}$ keV band (Kuntz et al. 2015).

Results from the more recent XMM-Newton and Suzaku missions have been mixed, consistent with the fact that these missions are primarily sensitive to O^{6+} and O^{7+} line emissions and the suggestion that the abundances of these species exhibit large variations that are uncorrelated with solar wind fluxes. Snowden et al. (2004) reported a case study in which the line of sight of the high altitude (apogee $\sim 114{,}000$ km) XMM-Newton spacecraft is thought to have passed through the subsolar magnetosheath. As expected, the intensity of 0.52–0.75 keV emissions (O^{6+} and O^{7+} lines) decreased midway through the observations in response to a decrease in solar wind flux while the O^{7+}/O^{6+} abundance ratio and 2–8 keV emissions remained nearly constant. However, neither of the intensities increased *earlier* in the observations when the solar wind flux and O^{7+}/O^{6+} abundance ratio increased. Collier et al. (2005b) explained the matches and mis-matches between the X-ray light curve and the solar wind flux with a tilted front model of the solar wind interacting with the Earth's magnetosheath.

Whittaker et al. (2016) reported that correlations between observed and modeled integrated line-of-sight intensities of soft X-rays with energies from 0.5 to 0.7 keV range from poor to excellent. Emissions in this band do not track the observed flux of solar wind protons, likely due to the large variation in the densities of the O^{7+} and O^{8+} ions (quantities that quite often are not available with sufficient accuracy or temporal resolution from solar wind monitor satellites).

However, there are a number of case studies demonstrating temporal correlations for emission thought to arise close to the Earth. Fujimoto et al. (2007) reported correlations between low altitude (~ 570 km) Suzaku spacecraft observations of 0.3–2 keV soft X-ray emissions and solar wind fluxes on time scales of about half a day when the line-of-sight passed through the northern cusp. Snowden et al. (2009a) reported a second case study of XMM-Newton observations in which O^{6+} emissions at 0.56 keV were weakly correlated with solar wind ion fluxes. On the other hand, Ezoe et al. (2010) reported an example in which Suzaku observations of 0.56 keV O^{6+} emissions tracked a transient increase in solar wind fluxes. Ishikawa et al. (2013) reported Suzaku observations indicating an increase in soft X-rays from charge exchange at 0.5–0.7 keV in conjunction with the arrival of enhanced solar wind fluxes during a geomagnetic storm, despite the fact that the line of sight of the spacecraft did not pass either through the subsolar magnetosheath or the cusps.

Examining XMM-Newton observations of the O^{6+} and O^{7+} lines on a statistical basis, Henley and Shelton (2010) concluded that there is no universal association between enhanced charge exchange emissions and solar wind flux enhancements. However, it should be noted that, due to XMM-Newton observing constraints, their compendium of observations were primarily through the flanks of the magnetosheath, where the charge-exchange emission will be dominated by emission from the local heliosphere. Carter et al. (2011) compared a large set of XMM-Newton soft X-ray light curves with corresponding variations in the solar wind flux. A scatter plot of soft X-ray intensities at energies of 0.5–0.7 keV *versus* solar wind ion fluxes showed only weak correlation. On a statistical basis XMM-Newton observations indicate a greater likelihood of observing soft X-ray emissions due to charge exchange at solar maximum than solar minimum (Carter et al. 2011).

From the results reported in this section, we conclude that solar wind fluxes are better correlated with band-integrated soft X-ray emissions in the $\frac{1}{4}$ keV band than the higher energy emission which is dominated by a limited number of lines.

6.2 Spatial Variability at Earth

This section first considers what information can be gained from low altitude spacecraft about the spatial variation of soft X-ray emissions generated by charge exchange in the vicinity of Earth, and then turns to observations by spacecraft with much higher apogees.

ROSAT and Suzaku observations were constrained to viewing directions $\pm 15°$ and $\pm 20°$, respectively, from the perpendicular to the Earth/Sun line. Because of this limitation and

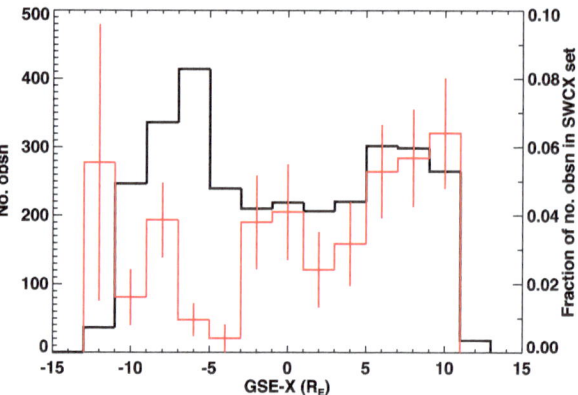

Fig. 38 XMM-Newton observations of solar wind charge exchange emissions (Carter et al. 2011). The black curve shows the total number of observations transverse to the Sun-Earth line as a function of position along the Sun-Earth line. The red curve shows the fraction of observations exhibiting enhanced emissions in lines associated with solar wind charge exchange

their low-altitude (< 600 km) orbits, the two spacecraft could only provide limited information concerning the spatial variations of soft X-ray emissions in the vicinity of Earth. With hindsight, it is now clear that ROSAT observations of soft X-rays from the dark side of the Moon (Schmitt et al. 1991), at intensities similar to those by which off-Moon fluxes exceeded those of the LTE-cleaned survey data in the same direction (Snowden et al. 1997), should be interpreted in terms of near-Earth solar wind charge exchange (Freyberg 1998). Indeed, Wargelin et al. (2004) have attributed more recent Chandra observations of strongly time varying O^{6+} and weaker O^{7+} emissions from the direction of the Moon in terms of charge exchange in Earth's vicinity.

As previously noted, Fujimoto et al. (2007) attributed low altitude Suzaku observations of enhanced emissions along a line of sight passing through the northern cusp to charge exchange occurring at distances between 2 and 8 R_E from Earth within this region. As discussed by Kuntz et al. (2015), the ROSAT survey removed short term (~ 10 minute long) enhancements in soft X-ray intensities near the Earth's magnetic poles, attributing them to auroral emissions. Some of these may have instead resulted from charge exchange in the cusps. Finally, within the limitations of the look directions nearly perpendicular to the Sun-Earth line available to low-altitude Suzaku, Ishikawa (2013) could find no evidence for any enhancement in soft X-ray emissions towards the dayside.

Global views of the solar wind-magnetosphere interaction require vantage points far outside the magnetosphere. Chandra and XMM-Newton are two missions whose orbits afford such views. Examining a set of XMM-Newton observations, Kuntz and Snowden (2008) concluded that the strongest 0.4 to 0.8 keV solar wind charge exchange emissions are usually observed on lines of sight that pass tangentially through the magnetosheath near the subsolar point. Nevertheless, strong emissions can still be observed on lines of sight through the flank magnetosheath when solar wind fluxes are particularly strong. Henley and Shelton (2010) also employed XMM-Newton observations but reported no tendency for the flux of soft X-rays corresponding to the O^{6+} and O^{7+} lines at 0.56 and 0.65 keV to increase when lines of sight pass through the subsolar region of the magnetosheath. In direct contrast, and as illustrated in Fig. 38, Carter et al. (2011) reported results from an extensive survey of XMM-Newton observations in the 0.5 to 0.7 keV band that indicated a strong preference for emissions to be observed when viewing through the subsolar magnetosheath. Kuntz et al. (2015) suggested that unsuccessful efforts to associate enhanced emissions with lines of sight passing through the subsolar magnetosheath probably resulted from the use of outdated static models to determine the magnetopause location. The XMM-Newton observations studied by Whittaker et al. (2016), of which some were correlated with modeled

Table 8 Observed solar wind charge exchange soft X-ray emission lines

Emission lines	Observatory	Author
C^{5+} O^{6+} O^{7+} Ne^{8+} Mg^{10+}	XMM-Newton	Snowden et al. (2004), Collier et al. (2005b)
C^{5+} O^{7+} Ne^{9+} Mg^{10+}	Suzaku	Fujimoto et al. (2007)
O^{7+} Ne^{8+} Ne^{9+} Mg^{10+} Mg^{11+} Al^{12+} Si^{12+} Si^{13+}	XMM-Newton	Carter and Sembay (2008), Carter et al. (2010)
O^{6+} O^{7+} Ne^{8+} Ne^{9+} Mg^{10+} Mg^{11+}	Suzaku	Bautz et al. (2009)
O^{6+}	XMM-Newton	Snowden et al. (2009a)
O^{6+} O^{7+}	XMM-Newton	Henley and Shelton (2010, 2012)
O^{6+}	Suzaku	Ezoe et al. (2010)
C^{4+} C^{5+} N^{6+} O^{6+} O^{7+}	Suzaku	Ezoe et al. (2011)
C^{4+} C^{5+} N^{6+} O^{6+} O^{7+}	Suzaku	Ishikawa et al. (2013)

See Fig. 15 for energies and intensities

emission while others were not, were selected to have lines of sight passing through the dayside magnetosheath. Cases with poor correlation may result from difficulties in locating the magnetopause. Since different MHD models predict different magnetopause locations, it would not be surprising if the comparisons of models with observations are sometimes less than satisfactory. Whittaker et al. (2016) also reported evidence for greater emissions in the dawn than dusk magnetosheath, perhaps suggesting an asymmetry in magnetosheath plasma densities.

6.3 Line Emissions Near Earth

In contrast to early missions like ROSAT, which observed soft X-rays in broad energy bands using either proportional counter or microchannel plate detectors, recent missions employ charge-coupled device (CCD) cameras to provide spectral resolution. These observations make it possible to identify some of the specific soft X-ray lines corresponding to solar wind charge exchange interactions as enhancements above the cosmic background. Table 8 summarizes the lines identified to date in observations of near-Earth soft X-ray charge exchange. In the absence of high spectral resolution observations, individual lines have not yet been identified at energies below 0.3 keV.

Dennerl et al. (1997) proposed using observations of line emissions to remotely sense differences in solar wind composition, i.e., between fast and slow speed streams and coronal mass ejections. Smith et al. (2005) subsequently interpreted Chandra observations of enhanced O^{7+} to O^{6+} emission ratios as evidence for a coronal mass ejection interacting with neutral gas in interplanetary space. Carter and Sembay (2008) reported that XMM-Newton observed greatly enhanced emissions corresponding to O^{7+} during the passage of a coronal mass ejection. Figure 39 shows how contributions from the species-dependent solar wind charge exchange line emissions can be combined with emissions from the non-variable diffuse sky and contributions from energetic protons to fit a background-subtracted and flare-cleaned XMM-Newton spectrum observed during a coronal mass emission (Carter et al. 2010). Ezoe et al. (2011) reported enhanced O^{6+} and O^{7+} emissions during a geomagnetic storm. Carter et al. (2011) employed ratios of Mg^{10+} to O^{6+} and O^{7+} to O^{6+} emissions to identify three candidate coronal mass ejections.

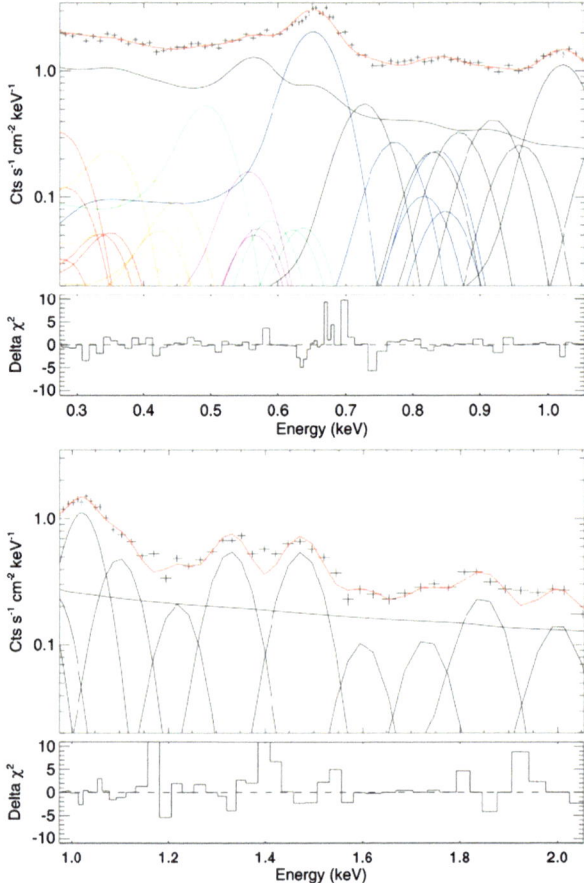

Fig. 39 A fit to the XMM-Newton observations of the soft X-ray spectrum seen during Obs301, a 48100s long interval from 1941 UT on 2001 October 21 to 0906 UT on 2001 October 22, at a time when the soft X-ray instrument was looking towards target 1Lynx.3a_SE (right ascension 08 h 49 m 06 s, declination $+44°51'24''$) through the dawn flank magnetosheath during the passage of a CME with solar wind proton number fluxes ranging from 3 to 28×10^8 cm^{-2} s^{-1} (Carter et al. 2010). The integrated background and solar flare related emissions have been subtracted. *Upper Panel*: The spectrum from 0.275 to 1.055 keV. *Lower Panel*: The spectrum from 0.975 to 2.055 keV. The sum of the non-variable sky and variable soft proton components is the continuous line in black. The lines due to C, N, and O are color coded: C^{4+} (red), C^{5+} (orange), N^{5+} (yellow), N^{6+} (green), O^{6+} (purple), and O^{7+} (blue). Heavier elements are in black. The residual at 1.4 keV may be due to incomplete background subtraction at the energy of the strong Al Kα instrumental line

With accurate magnetohydrodynamic models describing the Earth's plasma environment, it should also be possible to infer exospheric neutral densities from soft X-ray observations. Here observations suggest that models underestimate outer exosphere densities. Inspecting case and statistical studies of Suzaku observations, both Ezoe et al. (2011) and Ishikawa (2013) concluded that observed emissions require greater neutral densities beyond 10 R_E than current exospheric models predict.

Alternatively, with accurate magnetohydrodynamic and exospheric neutral density models, it should be possible to determine band integrated soft X-ray cross-sections. Kuntz et al.

(2015) compared ROSAT observations with predicted line of sight emission rates to obtain a value of 3.8×10^{-20} counts deg^{-2} cm^4 for the $\frac{1}{4}$ keV band. Noting that this value is 1.8 to 4.5 times higher than values derived from limited atomic data (Robertson et al. 2009b; Koutroumpa et al. 2009a, and see Sect. 3.1.3), they suggested that the atomic data may be missing a large number of faint lines. In summary, observations from past and present missions suggest that the intensity of soft X-ray emissions from charge exchange far exceeds predicted levels.

6.4 Comets

ROSAT observations of comet Hyakutake (Lisse et al. 1996) led to the discovery of soft X-rays resulting from charge exchange (Cravens 1997). Since then, we have learned that all comets are X-ray sources with luminosities as high as 1 GW and emissions extending over hundreds of thousands of kilometers (Cravens 2002; Dennerl et al. 1997; Krasnopolsky et al. 2004). Although the densities of solar X-ray photons greatly exceed those of solar wind heavy ions, the far greater cross-sections for heavy ion charge exchange ($\sim 10^{-15}$ cm^2) than X-ray scattering ($< 10^{-18}$ cm^2) ensure that charge exchange emissions predominate at objects with low density gases spread over a large volumes, such as comets and planetary magnetosheaths. Consequently the cometary spectra measured by Lisse et al. (2001) and others exhibit the transition lines expected for charge exchange with high-charge state solar wind species.

Wegmann et al. (2004) demonstrated that gas production rates determined from soft X-ray measurements correspond well to those calculated by other methods. They went on to show that integrated soft X-ray emissions track variations in the solar wind density, and attributed those changes in emission that did not track variations in the solar wind density to variations in the solar wind composition. Lisse et al. (2005) used Chandra observations to identify a solar wind composition consistent with low densities and high effective temperatures in a high-speed post-shock bubble with charge state compositions intermediate between the fast and slow solar winds passing by the vicinity of Comet 2/Encke. Lisse et al. (2007) associated peaks in soft X-ray emissions with the arrival of increases in the solar wind flux. Since comets occur relatively frequently, observations of their soft X-ray emissions can be used to monitor changes in solar wind composition over the solar cycle (Dennerl et al. 2012) and to probe the difference between the equatorial and polar solar wind flows (Bodewits et al. 2007).

As seen in Fig. 40, Wegmann and Dennerl (2005) used XMM-Newton soft X-ray observations to construct spatial images of the dynamic shock position and shape at comet C/2000 W1 (LINEAR). Consistent with results from *in situ* observations, they concluded that the $\sim 40,000$ km shock transition is far broader than the ~ 100 km transition seen at Earth (Russell et al. 1982a), with distinct inner and outer boundaries. As expected, the outer boundary is well approximated by a parabola. They attributed north/south asymmetries in sheath brightness and inferred shock thickness to effects associated with the orientation of the interplanetary magnetic field. In particular, they maintained that the quasi-parallel bow shock is broader than the quasi-perpendicular bow shock, and that thermal velocities and therefore X-ray intensities are greater in the magnetosheath behind the quasi-parallel bow shock. Note that Wing et al. (2005) have also invoked greater temperatures in the quasi-parallel magnetosheath than in the quasi-perpendicular magnetosheath to explain temperature patterns in the terrestrial magnetotail plasma sheet. Consequently there are good reasons to expect similar asymmetries in soft X-ray observations of the Earth's magnetosheath.

Finally, observations of soft X-rays from comets can be used to remotely sense solar wind conditions. Neugebauer et al. (2000) noted that both ROSAT and the Extreme Ultraviolet

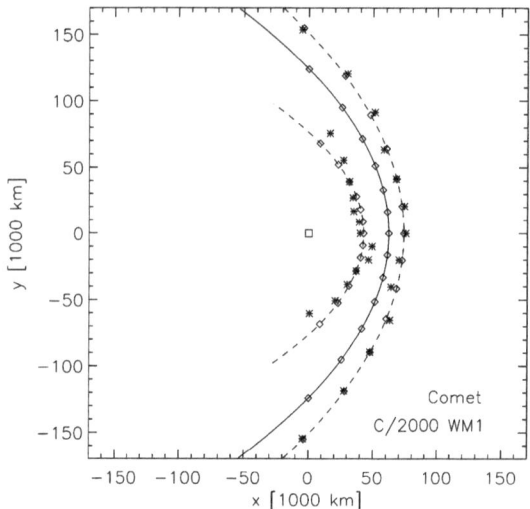

Fig. 40 The predicted location of a thin bow shock (solid line), locations of the inner and outer edges of the observed thick bow shock derived from XMM-Newton observations (asterisks), and parabolic fits to the observed locations (diamonds and dashed curves) for Comet C/2000 WM1 (LINEAR) (Wegmann and Dennerl 2005). The square at the center of the figure marks the position of the cometary nucleus

Explorer spacecraft observed time varying soft X-ray emissions from comet C/Hyakutake 1996 B2. They showed that much of this variability can be explained in terms of the varying solar wind. However, they cautioned that emission strengths also depend upon solar wind composition and comet outgassing. Kharchenko and Dalgarno (2001) interpreted observations of emissions from comets Levy and Hale-Bopp as evidence for their interaction with a slow solar wind.

6.5 Mars and Venus

Past work has already demonstrated the feasibility of imaging the plasma environments of Mars and Venus (Dennerl et al. 2012). As predicted by Cravens (2000b) and Krasnopolsky (2000), and modeled by Holmström et al. (2001), Gunell et al. (2004, 2005), and Koutroumpa et al. (2012) charge exchange at both planets results in the emission of soft X-rays. Chandra detected only a weak soft X-ray halo around Mars (Dennerl 2002). Figure 41 shows Chandra observations of this X-ray halo. However, Dennerl (2006) and Dennerl et al. (2006) then used XMM-Newton observations to demonstrate that fluorescent scattering dominates observations from the disk itself, but that emissions from charge exchange dominate at greater distances. Charge-exchange intensities vary with time in a manner different that those of solar X-rays in response to fluctuations in solar wind parameters and exospheric densities. Although emissions were strong when observed by XMM-Newton in 2003 at a time of high solar activity (Dennerl et al. 2006), they were insignificant during Suzaku observations in 2008 (Ishikawa et al. 2011), enabling the latter authors to place an upper limit on exospheric neutral densities during solar minimum. When observed, emissions diminish with radial distance from 1–3 R_M, but can be detected as far out as 8 R_M, well beyond the distances where the models listed above predict significant emissions. Images of the Martian emissions provide clear evidence for a bow shock-like shape and intensities that increase dramatically with increasing solar activity.

Contamination from scattered solar X-rays precluded identification of charge exchange emissions from Venus in Chandra observations during solar maximum (Dennerl et al. 2002). However, Chandra observations during solar minimum demonstrated that the dayside magnetosheath of Venus also emits X-rays that outline the locations of the bow shock and

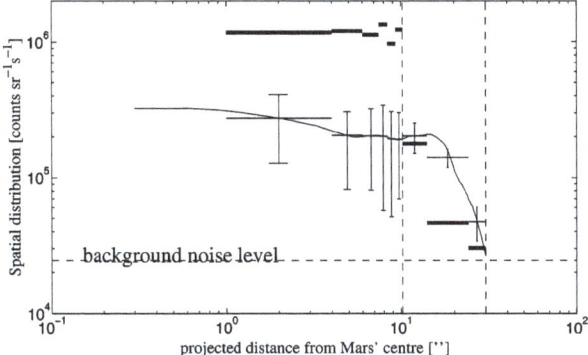

Fig. 41 The radial distribution of Chandra X-ray observations at and around Mars in units of counts sr^{-1} s^{-1} (Gunell et al. 2004). The thin solid curve shows results from a charge exchange simulation. The thick horizontal lines show observations. Fluorescent scattering of solar X-rays causes count rates over the disk to greatly exceed those predicted for charge exchange (projected distance $< 10''$), but there is reasonable agreement in the halo extending from $10''$ to $25''$ from the center of the planet

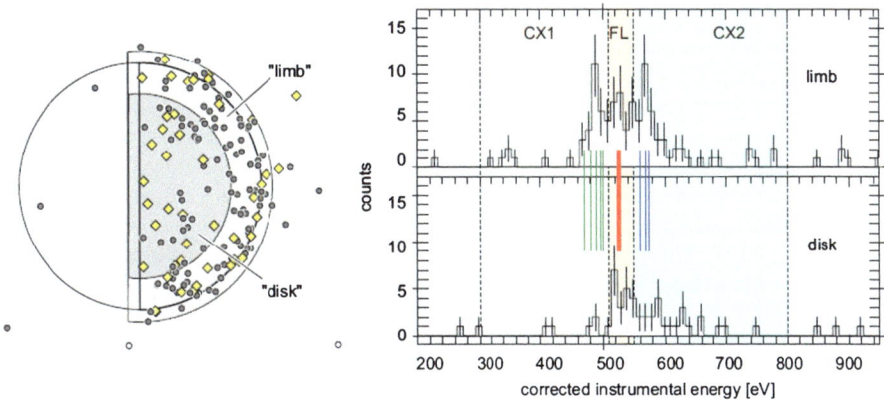

Fig. 42 Chandra observations of soft X-ray spectra at Venus obtained on 2006 March 27 (Dennerl 2008). *Left Panel*: The distribution of the X-ray photons in the corrected instrumental energy range from 0.3 to 0.8 keV. Bright diamonds mark photons in the fluorescence band, while dark circles mark those in the charge exchange bands CX1 and CX2. The extraction regions for the limb and disk spectra are superimposed in light and dark gray, respectively. The circle indicates the geometric size of Venus. *Right Panel*: The X-ray spectra for the limb (upper) and the disk region (lower). The spectrum in the limb region is dominated by two emission lines in the CX1 and CX2 bands. These lines are almost absent in the disk spectrum, which is dominated by emission in the FL band. Vertical lines mark the energies of lines expected to characterize charge exchange and fluorescence

ionopause (Dennerl 2008, 2010). Observed emissions matched the predictions of Gunell et al. (2007) rather well, indicating that we understand charge exchange better at Venus than Mars. Figure 42 compares soft X-ray spectra observed over the dusk and limbs of Venus.

Soft X-ray emissions can not only be used to derive information concerning plasma structures in the vicinity of Venus and Mars, but also the shape of their exospheres. Regions with enhanced neutral densities emit more soft X-rays. In contrast to results expected for the symmetric exospheres predicted by models (Holmström et al. 2001), soft X-ray emissions from the Martian magnetosheath peak over the poles (Dennerl et al. 2006), indicating that the Martian exosphere is not spherically symmetric. Because modeling shows that asymmetries in the exobase propagate to large distances from the planet, global images can be used to detect any seasonal variations in the Martian atmosphere (Holmström 2006).

6.6 The Moon

On 29 June 1990, ROSAT made the first soft X-ray image of the Moon (Schmitt et al. 1991). Soft X-rays were observed even on the nightside of the Moon where scattering is too inefficient to explain their presence. Instead, the ROSAT observations were interpreted in terms of continuum emission from solar wind electrons impinging on the lunar surface. In 2001, Chandra detected soft X-ray emission from O^{+6} and O^{+7} in two sets of observations on the dark side of the moon (Wargelin et al. 2004). Solar wind charge exchange emissions occurring between the Earth and the Moon in the magnetosheath explain both the Chandra and ROSAT observations.

Interactions between the solar wind and the tenuous lunar exosphere (Stern 1999), which lies exposed to the solar wind during about two thirds of the lunar orbit, also generate soft X-rays. Robertson et al. (2009a) modeled the soft X-ray intensities produced by charge exchange at the Moon and predicted them to be about 10 keV cm^{-2} s^{-1} sr^{-1}, comparable to the diffuse cosmic X-ray background (Lumb et al. 2002).

Collier et al. (2014) presented observations of limb brightening in the ROSAT lunar data consistent with the expected signal from solar wind charge exchange with the lunar exosphere and compared these observations to the predictions of hybrid simulations for solar wind ion access to regions behind the terminator of the Moon (Halekas et al. 2005; Farrell et al. 2008; Fatemi et al. 2012) and models of the lunar exosphere (Sarantos et al. 2012; Tenishev et al. 2013). Of particular note is that the soft X-rays produced by charge exchange can provide diagnostics on the entire neutral exospheric density including all species, whereas other techniques are spectroscopic in nature and observe particular lines, for example Na (Potter et al. 2000). Thus, the soft X-ray observations provide a LOS measure of the radial profile of the entire lunar exosphere.

Soft X-ray observations of the lunar exosphere complement and validate model predictions for the dominant contributors to the exospheric column density and will address a wide variety of lunar science topics beyond those discussed by Collier et al. (2014). Plasma structures in the vicinity of the Moon include a low density wake (Lyon et al. 1967; Zhang et al. 2014) and mini-magnetospheres above magnetic anomalies (Wieser et al. 2010) that could be imaged globally using soft X-ray emission.

Soft X-ray imagers can monitor the relationship between the time-variable solar wind flux and composition and the lunar exosphere remotely, similar to what Lunar Reconnaissance Orbiter observed *in situ* (Feldman et al. 2012) by showing that the surface He density exhibits variations responding to changes in the solar wind alpha flux. Soft X-ray observations can also be used to study the behavior of the exosphere as the Moon moves in and out of the terrestrial magnetotail, which turns off and on the effect of the solar wind driver on the exosphere.

Soft X-ray observations may be used to determine the constituents of the lunar atmosphere over the poles by comparison with the local time dependencies predicted by models. This information can be used to estimate the production rates and polar cold trapping efficiencies for the non-condensible species on the cold lunar nightside.

6.7 The Heliosphere

Soft X-rays emitted from the heliosphere are omnipresent in X-ray observations, since they are generated over tens of astronomical units around Earth-bound observatories and cannot

Fig. 43 Observation geometry for the DLX sounding rocket flight and the ROSAT survey observations for the He focusing cone observations (Fig. 1 of Galeazzi et al. 2014)

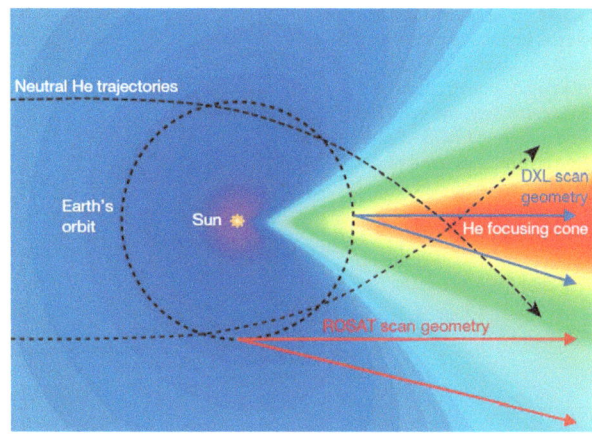

be avoided by carefully chosen observation geometries as in the case of soft X-rays emitted in Earth's vicinity. Because their spectrum resembles that for thermal emissions from the local hot bubble and their signal integrated over heliospheric distances smooths out both spatial and temporal variations, it has proven quite challenging to isolate the signal of heliospheric soft X-rays from those emitted within the Local Hot Bubble. The first attempt to detect SWCX emission from the heliosphere (X-ray observations of the helium focusing cone) was reported by Snowden et al. (2009b).

The Diffuse X-ray emission from the Local galaxy (DXL) sounding rocket flight was designed to isolate the spatial variation of soft X-ray emissions from the He focusing cone (Galeazzi et al. 2011, 2012, 2014; Uprety et al. 2016). The look direction of the local midnight launch on 2012 December 13 from the White Sands Missile Range in New Mexico avoided soft X-rays from charge exchange in the vicinity of the Earth's magnetosphere and pointed antisunward through the maximum emission region of the heliospheric He focusing cone, a direction towards the galactic anti-center which avoids bright galactic soft X-ray sources.

As illustrated in Fig. 43, the strength of the soft X-ray emissions from the He focusing cone was identified when DXL observations through the He focusing cone were compared with previous RASS observations of the same region in the sky along a line of sight parallel to but not through the He focusing cone. Parallax effects and the specific observation geometry had precluded all but a fraction of the emissions from the He focusing cone from being observed by the RASS (see Fig. 10 of Koutroumpa et al. 2009a; Lallement et al. 2004b). Simultaneous fits to the DXL and RASS data shown in Fig. 44, together with models predicting differences between the soft X-ray signals between the two missions, allowed Galeazzi et al. (2014) to put tight constraints on the heliospheric SWCX signal. They conclusively demonstrated that $\sim 40\%$ of the total soft X-ray diffuse emission in the galactic plane at energies near 250 eV band originates from the He focusing cone and not the Local Hot Bubble surrounding the heliosphere.

7 Imaging Technologies

The theoretical predictions and simulation results presented in Sects. 3 and 5 demonstrate that it should be possible to image the location, motion, and structure of the Earth's bow

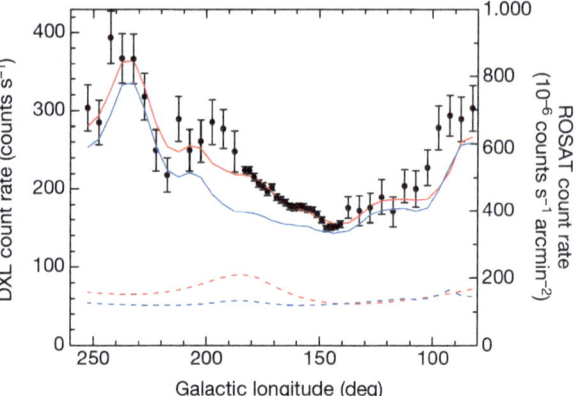

Fig. 44 DXL and ROSAT count rates as a function of position on the sky (Galactic longitude). The data points and errors show the DXL count rate, the solid blue curve shows the ROSAT count rate, the blue and red dashed curves show the model heliospheric SWCX contributions to the ROSAT and DXL count rates, respectively. The solid red curve is the best fit to the DXL total count rate (Fig. 4 of Galeazzi et al. 2014)

shock, magnetosheath, magnetopause, and cusps in soft X-rays. The review of past observations presented in Sect. 6.1 confirms that these signatures have been routinely detected by astrophysical telescopes with narrow fields-of-view. However, addressing the scientific topics discussed in Sect. 2 requires global images of the dayside solar wind-magnetosphere interaction. This section describes existing and planned instrumentation that will enable researchers to construct these global images. It reports the results of simulations designed to illustrate the images which current technology can achieve.

7.1 General Considerations

This section explores the technological requirements for imaging the magnetosheath and cusps. Although instruments that scan across the region of interest can build up global images, scanning a region in two dimensions with a non-imaging or narrow FOV instrument reduces observing efficiency and requires additional spacecraft resources compared to instruments capable of viewing the entire FOV simultaneously. These overheads could be quite extreme if the requirement is to detect spatial variability on short timescales. For this reason we will not consider scanning instruments further.

The dayside magnetosheath and its boundaries have dimensions of $\sim$ 10–20 R_E and exhibit substructures on various scale sizes down to a fraction of an R_E. Depending on the location of the imager the scientific requirement for the FOV is at least some tens of degrees. The angular resolution requirements are quite modest and are at minimum around 30 arc minutes for anticipated viewing distances. The magnetosheath structure is known to vary dynamically in response to the solar wind. This variability operates on timescales as short as 1–2 minutes, hence we have the joint requirement of providing a large FOV and sufficient angular resolution and flux sensitivity to detect potentially short timescale changes within our global images.

In addition to the constraints imposed by the geometrical size of the magnetosheath, any instrument design will be informed by the orbit of the imager and also the fact that both the bright Earth and the Sun will impose strict avoidance constraints and have physical implications on the instrument design with regards to stray light baffle sizes. Solar wind charge exchange produces a spectrum of photons in line emissions from extreme ultraviolet (EUV) to soft X-rays (around 10 eV to 2.2 keV). For most solar wind conditions, the predicted integrated SWCX emissivity in the EUV (i.e., below $\sim$ 100 eV) is comparable to the integrated emissivity in soft X-rays (i.e., above $\sim$ 100 eV). Although the strongest individual

SWCX line is the O^{6+} line at 12 eV, observations at such long wavelengths in the EUV are essentially impossible due to the extremely strong and extended emission from geocoronal Lyα (1216 Å = 10.2 eV). In addition, the Earth's plasmasphere extends several Earth radii outward, exhibiting bright emission due to resonant scattering of the input solar EUV spectrum by O II and He II ions at 834 Å (14.9 eV) and 304 Å (40.8 eV), respectively. Planet B EUV scanner observations indicate that the surface brightness of the emission at 40.8 eV decreases from $\sim$ 5 Rayleigh at $L = 5$ to 0.5 Rayleigh across the region where the generally sharp plasmapause lies (Nakamura et al. 2000).

Hence global imaging of solar wind charge exchange must be restricted to energies above $\sim$ 50 eV. Imagers that operate in the EUV band can use normal incidence (NI) optics with a multilayer coating designed to produce a suitable reflectance. Regularly spaced multicoatings have, however, an extremely narrow energy bandpass at short wavelengths (typically $\sim$ 20 Å) which is useful for isolating strong lines for some experiments (such as imagers designed to observe Earth's plasmasphere) but much less useful for imaging solar wind charge exchange. In addition, the reflectivity has a narrow response as a function of the angle of incidence which limits the performance for wide field of view EUV instruments. Theoretically the spectral and angular bandpass of a multilayer coating can be broadened by varying the depths of the individual layers (van Loevezijn et al. 1996; Wang et al. 2000) at the expense of peak reflectivity, however, no flight instrument has yet used such a coating.

For X-rays, the critical angle for reflectance of a surface is typically $\sim$ 1°. Hence all soft X-ray imagers use grazing incidence (GI) optics of one form or another. NI optics are theoretically more mass efficient than GI optics for a given aperture area and FOV (e.g., Gorenstein 2010). Nevertheless, GI X-ray instruments have a clear advantage over NI EUV instruments due to their much broader spectral response. This makes the science return of the instrument much more robust against variability in the SWCX spectrum and increases the potential secondary science objectives. Furthermore, it is technologically much easier to provide useful energy resolution for a detector system in X-rays than in the EUV. Observations of terrestrial solar wind charge exchange emissions by XMM-Newton and Suzaku using CCD detectors have shown that such instruments can directly provide charge state information on the heavy ion content of the solar wind (e.g., Carter et al. 2010; Ishikawa 2013). For these reasons, to date, all proposals for instruments designed to perform global imaging of terrestrial SWCX have been soft X-ray instruments.

7.2 Soft X-Ray Imaging Technologies

This section of the paper describes soft X-ray imaging technologies. It begins with a discussion of Wolter-type and lobster-eye optics used in the past and planned for future missions. It then turns to the topic of detector planes with a discussion of microchannel plates, charge-coupled devices, and active pixel sensors. This section concludes with a brief summary of existing and planned wide field-of-view soft X-ray imagers on the DXL, DXL-II, CuPID, SMILE, and other future missions.

7.2.1 Optical Elements

Grazing incidence optics in almost all flight X-ray telescopes to date have used a Wolter type I geometry consisting of two sets of nested shells with paraboloid and hyperboloid surfaces, or a conical approximation thereof. The angular resolution of these flight mirrors is easily much smaller than the typical minimum size ($\sim$ 30') of the magnetosphere structures that must be imaged. A consequence of the small critical angle for reflectance is that the

Fig. 45 *Left Panel*: The BepiColombo MIXS instrument with the MIXS-T Wolter type 1 optic on the left and the MIXS-C collimator optic on the right in the photograph. *Right Panel*: The channel structure of a square-packed MPO plate

theoretical FOV of such mirrors is at most $\sim 2°$ in radius in soft X-rays, so an imager viewing the Earth's magnetosheath using Wolter optics would need multiple modules to meet the FOV requirements. Mass requirements probably preclude such conventional nested shell optics for a small mission.

However, lightweight Wolter-type optics have been developed that rely on structures which are, to a significant extent, self supporting. These include microchannel plate (MCP) optics, equivalently named Micro Pore Optics (MPOs), manufactured by the Photonis Corporation in collaboration with the University of Leicester, UK. Manufacturing MPOs essentially consists of fusing together glass capillaries into plates containing an array of millions of square glass pores (Mutz et al. 2007). The capillaries initially have a glass cladding tube and a soluble glass core that is etched away to form the pore. The plates are then thermally slumped to the required radius of curvature.

The Mercury Imaging X-ray Spectrometer (MIXS, Fig. 45, left panel) instrument on ESA's Bepi-Colombo mission, which was scheduled for launch in 2018, employs MPOs (Martindale et al. 2009; Fraser et al. 2010). MIXS has two channels. In the MIXS-T optic the pores are radially packed and the plates are then stacked to form a conical approximation to the Wolter type I geometry. The MIXS-C optic has a single layer of four plates with square packed pores and the optic is employed as a collimator. MPO plates are typically around 4 cm × 4 cm in size, with 20 μm pores, sidewalls of 6 μm and an open area of $\sim 60\%$. Assembly of a complete optic requires mounting plates within a frame. Figure 45 (right panel) shows a micrograph picture of the channel structure of a square-packed MPO plate.

Ezoe et al. (2009) reported using micromachining technologies to etch the curvilinear micropore sidewalls through a thin silicon wafer for use as reflecting surfaces. As with the MPOs, two such wafers deformed to different curvatures of radii can be stacked to work in the Wolter geometry. The optics were fabricated by Tokyo Metropolitan University and ISAS/JAXA in collaboration with engineering researchers. The wafer size and thickness are typically 4 inch (i.e., 10 cm) and 300 μm respectively, while the pore width and walls between pores are both 20 μm. The largest possible diameter of the optic is 30 cm. The open area ratio is $\sim 30\%$, smaller than that for MPOs, but the larger wafer size relaxes assembly of the whole optics system.

Figure 46 shows an example of a 4-inch Wolter type-I optic. The optic covers a 4 degrees diameter field of view with a focal length of 25 cm. Although it has a conically approximated Wolter type-I geometry, the thinness of the wafer makes the effect negligible. The theoretical

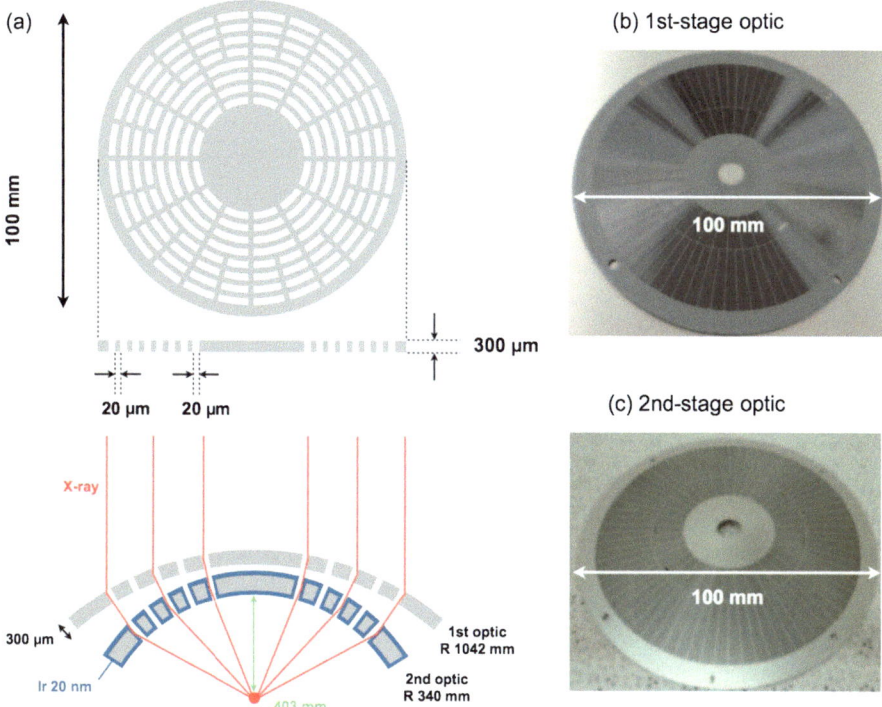

Fig. 46 *Left Panel*: Overview of the light-weight Wolter type 1 optic based on micromachining technologies. *Right Panel*: Examples of the first and second stage optics (Ogawa et al. 2017)

Fig. 47 Micrograph showing a micropore for the micromachined Wolter type 1 X-ray mirror (Ogawa et al. 2017)

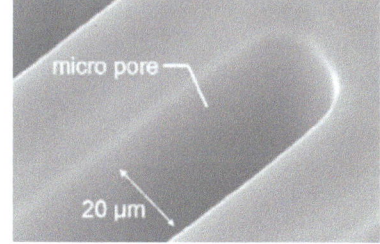

limit on the angular resolution comes from X-ray diffraction due to the narrow pores, which is estimated as 13″ at 1 keV and 20 μm pore width (Fig. 47). To date, X-ray imaging with the Wolter type-I sample has been demonstrated, while the angular resolution was of the order of 10′ (Ogawa et al. 2017). Efforts are being made to increase resolution by improving the fabrication processes.

Figure 48 shows the effective area and grasp (the product of the effective area and solid angle, $A \times \Omega$) of the 4 inch Wolter type-I optic made from 300 mm thick wafers. Grasp is a measure of sensitivity to diffuse X-rays extended over the field of view, as is the case with solar wind charge exchange emissions. Similar to the MPO, both parameters depend strongly on the wafer coating and the mirror surface roughness. Nickel- and Iridium-coated silicon optics achieve greater effective areas than purely silicon optics at all energies except the at the L-shell absorption edge of Nickel around 1 keV. Nickel provides the greatest

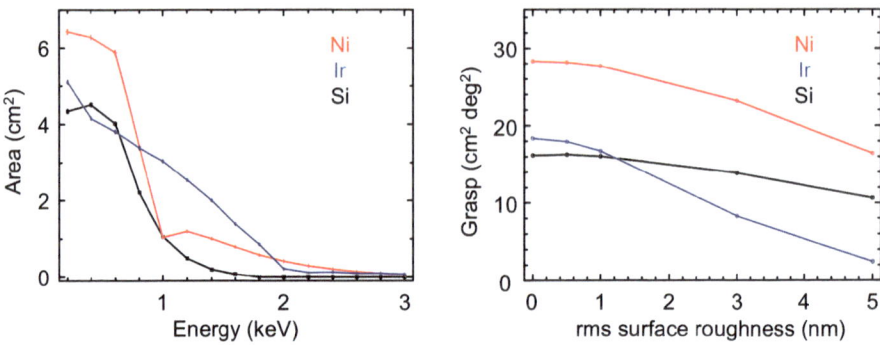

Fig. 48 *Left Panel*: The on-axis effective area per optic of a 4 inch optic with focal length of 25 cm as a function of energy. Different colors correspond to different coatings. Mirror surface roughness is fixed at 1 nm rms. *Right Panel*: The grasp at 600 eV as a function of the mirror surface roughness in rms (Mitsuishi et al. 2016)

effective areas for the energies below 1 keV which are of greatest interest. From recent experiments, a mirror surface roughness of 1 nm or less is achievable. Because of the large field of view, the ~ 1000–2000 mm^2 deg^2 grasp for an rms roughness of 1 nm is comparable to that of the larger telescopes on board Chandra or Suzaku. The 4 inch Wolter type-I optic is intended for future Japanese planetary missions that will employ X-ray imaging to explore the magnetospheres of Earth and Jupiter (Ezoe et al. 2013). A single optic or an array will be used to take high temporal and spatial resolution X-ray images and spectra in the near future.

An alternative to the Wolter geometry is the Lobster-eye geometry, so-called because of its similarity to the structure of a crustacean's eye (Angel 1979). The basis of the geometry is that the reflecting surfaces are arranged orthogonal to a spherical surface. Optics with this property have been developed at Leicester University in collaboration with the Photonis Corporation using MPOs as noted earlier (Fraser et al. 1992). Parallel incident rays are brought to a focus on a spherical surface at a distance of half the radius of the curvature of the optic. The left panel of Fig. 49 shows a simplified picture of the focusing geometry of a Lobster-eye optic.

Lobster-eye optics have the properties that the vignetting function *and the size of the PSF* are uniform across most of the field of view which would not be the case of an array of Wolter type optic modules built to provide the same field of view. Their disadvantage is that the point-spread function (PSF) of such an optic is very complicated with the majority of the power of the PSF spread out at large angles. The right panel of Fig. 49 shows a ray-tracing simulation of the PSF from a typical MPO. The image is logarithmically scaled to show the off-axis structure. The square channels of the MPOs allow multiple modes of reflection with some fraction of the photons simply collimated. The central focused core of the PSF contains around 17% of the photons and has a relatively narrow full width half maximum of a few arc minutes (and is thus capable of identifying point sources in the image which would be a source of background in this application). The focused core is due to photons that have an odd number of reflections from one pair of roughly parallel walls and an odd number of reflections from the orthogonal pair of roughly parallel walls. The cross-structure in the image is due to photons with an odd number of total reflections. The central island comprises photons with no reflections (i.e., essentially photons collimated by the MPO). The islands at larger radii have an even total number of reflections but without the constraint of having equal numbers of reflections from orthogonal walls.

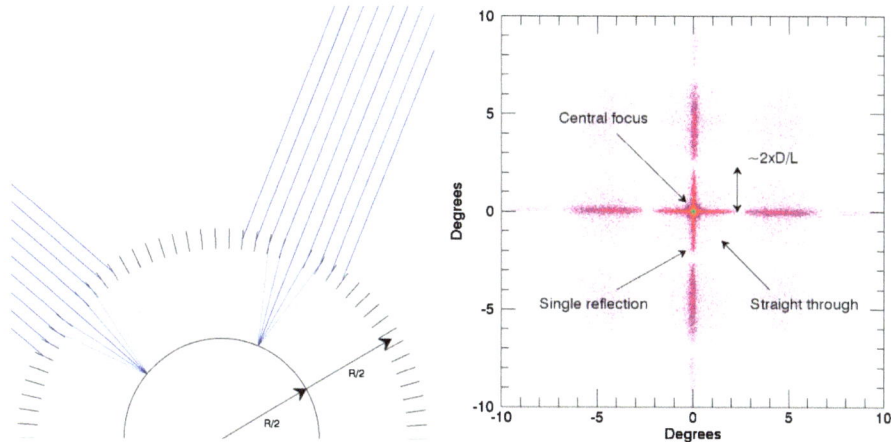

Fig. 49 *Left Panel*: Simplified view of the focusing geometry for a Lobster-eye optic. X-rays that make grazing incidences on the optics are focused only a detector plane at half the radius of curvature. *Right Panel*: Binned image for a micropore optics point spread function generated by a ray-tracing simulation. The image has been logarithmically scaled to emphasize the structure at large off-axis angles. Labels indicate the origin of various features in the image

Around 30% of the photons at 0.5 keV from a point source are recorded within a radius of 30′ from that source, which typifies the bin size that would be used to study diffuse emissions. Full extraction of the spatial information in the science images will require a well-calibrated PSF.

The left panel of Fig. 50 shows the simulated effective area of an MPO array. The MPO configuration is assumed to consist of Iridium-coated 500 mm focal length plates with an open area of 61% and pore length to diameter ratio of 50:1. The plot shows, as a function of energy, the total effective area (i.e., within 100% of the PSF) and the effective area within a 30′ radius (approximately three times the typical full-width at half-maximum, FWHM, of the PSF) around the PSF core. The optic is assumed to be mounted on a frame consisting of 38 × 38 mm apertures with 4 mm wide cross pieces. Figure 50 (right panel) shows the calculated Half-Energy Width (HEW) and FWHM of the PSF, also as a function of energy. The lower panel of Fig. 50 shows the encircled energy as a function of PSF radius, and indicates the HEW. The simulation software is based around a physical model of the MPOs and their intrinsic and manufacturing aberrations as outlined in Willingale et al. (2016).

The effective area and field of view of a given MPO depend jointly on its physical dimensions and the physical dimensions of the channels. For an on-axis source at infinity, the physical area of the optic contributing to the focussed core is the minimum of the detector area and $2 R_c \times a_{crit}$ where R_c is the radius of curvature, $a_{crit} \sim \arctan(2w/l)$, w is the channel width, and l is thickness of the optic. The size of the FOV is $\sim [D/R_c - 2a_{crit}]$ where D is the width of the optic. Figure 51 shows how these parameters vary as a function of focal length for a fixed size uncoated optic. For a given physical optic size (and therefore also a detector size of $\sim 1/2$ the optic size) there is a trade-off between effective area and FOV.

As for Wolter type optics (e.g., Fig. 48), the effective area of the MPO also depends on the optic coating. Figure 52 shows model effective areas extended to lower energies for different coatings along with the transmission of a typical optical light blocking filter. Also shown is a model SWCX spectrum. Assuming that the MCP QE is relatively flat between

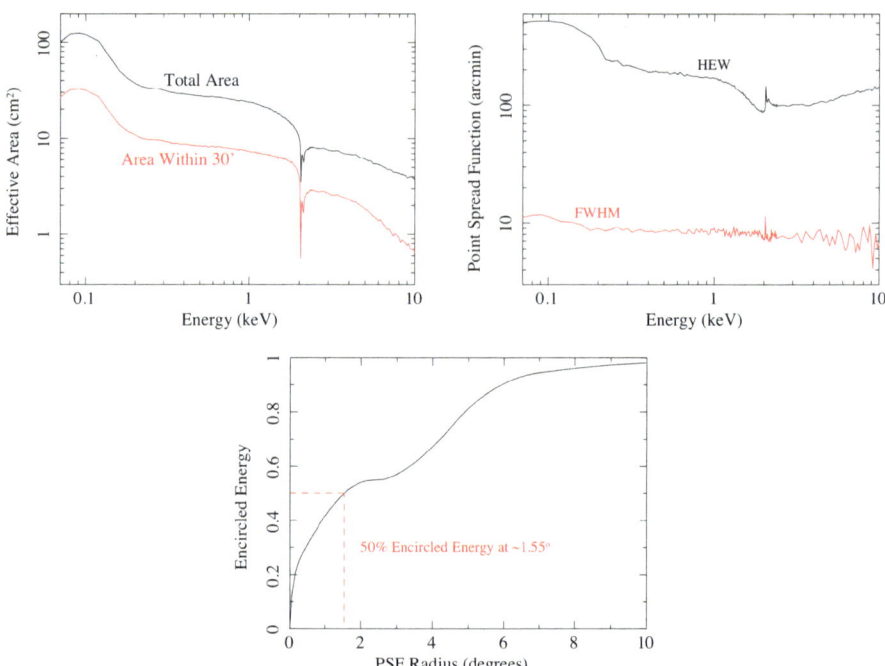

Fig. 50 *Left Panel*: The simulated effective area of an MPO array. The black curve shows the effective area for the full instrumental throughput (see text for details). The red curve show the effective area for the X-rays falling within a 30′ radius resolution element. *Right Panel*: The calculated HEW (black curve) and FWHM (red curve) of the PSF of the MPO, also as a function of energy. The curves show the effects of the broad wings but narrow core of the PSF. *Lower Panel*: The encircled energy fraction as a function of radius for the PSF

Fig. 51 The relation of the diameter of the FOV (solid red) and the active area (solid black) as a function of the focal length (half the radius of curvature) when the size of the device is held constant. The dashed red line is the α_{crit} for the device and is a typical value. Note that the *active area* is merely the surface area of the optic that can focus light parallel to the optical axis. The *effective area* is significantly smaller than the active area

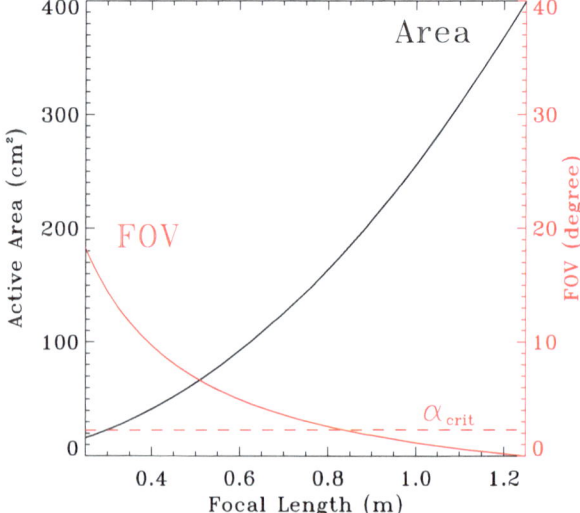

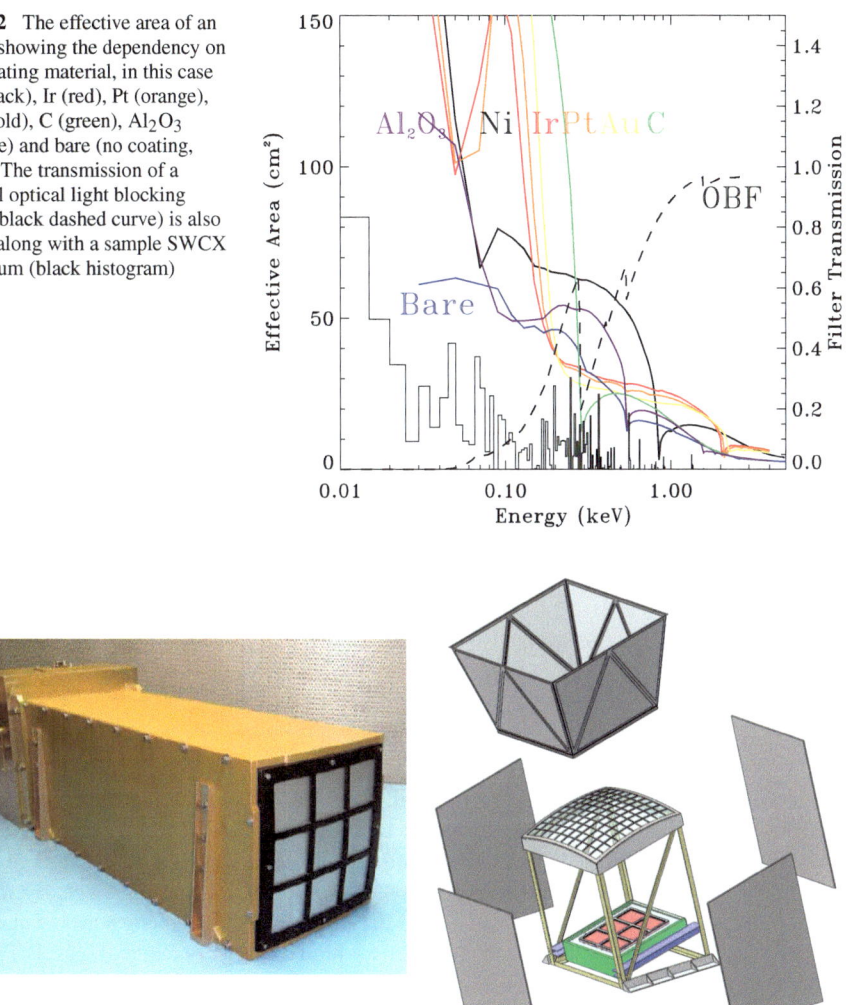

Fig. 52 The effective area of an MPO showing the dependency on the coating material, in this case Ni (black), Ir (red), Pt (orange), Au (gold), C (green), Al_2O_3 (purple) and bare (no coating, blue). The transmission of a typical optical light blocking filter (black dashed curve) is also show along with a sample SWCX spectrum (black histogram)

Fig. 53 *Left Panel*: STORM prototype X-ray imager with a 3×3 optics assembly at the front and an electronics assembly at the back (Collier et al. 2012). *Right Panel*: Complete module concept which has a 10×10 micropore optics array and a 3×2 MCP detector array. A baffle shields the optics from solar and terrestrial emissions

0.1 and 1.0 keV, the coating should be optimized to produce the best effective area at the energy with the greatest flux, roughly 0.1–0.5 keV. In this band Ir is little better than bare glass, while Ni provides a substantial improvement. At higher energies, 0.5–2.0 keV, Ir is a significant improvement to bare glass while Ni has a strong edge near several important emission lines.

The left panel of Fig. 53 shows a prototype Lobster-eye instrument using MPO plates with a focal length of 37.5 cm that has undergone laboratory testing (Collier et al. 2012), and has successfully been flown as a sounding rocket instrument (Thomas et al. 2013; Collier et al. 2015). It is the first instrument to detect X-rays from space using MPOs. The instrument has a mass of 7.4 kg. The optic assembly can take up to nine MPO plates although only two

Fig. 54 Comparison of quantum efficiency (QE) curves for various detectors. Black Curves: back-side illuminated CCDs, solid: XMM-Newton pn, dashed: Suzaku XIS1, dotted: Chandra ACIS-S3. Red curves: front-side illuminated, solid: XMM-Newton MOS2, dashed: Suzaku XIS2, dotted: Chandra ACIS-S2. Blue curves: MCPs, solid: Chandra HRC-I, dashed: Chandra HRC-I times 1.25 (the Chandra HRC had lower than expected QE)

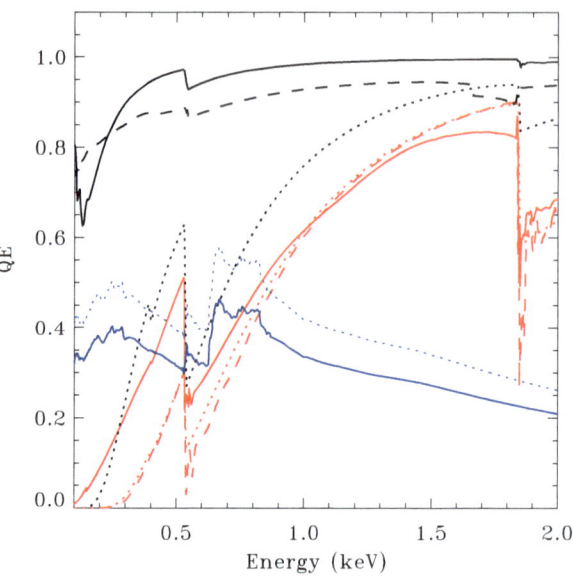

facets were filled on its first flight. The right panel of Fig. 53 shows a concept for an X-ray imager with a 10 × 10 facet MPO array and a 2x3 MCP detector array.

7.2.2 Detectors

One consequence of the optic geometry is that the size of the detector must be at least half the size of the optic. This requirement is a challenge which can be met in several ways: microchannel plate (MCP) detectors, silicon based detectors such as charge-coupled devices (CCDs), or variants of active pixel sensors (APS) such as complementary metal-oxide-semiconductor (CMOS) devices. Both MCPs and CCDs have a long heritage in solar-terrestrial and astrophysics applications and have a high technology readiness level (TRL).

CCDs are large format silicon detectors where the charge generated by an absorbed X-ray is localized within the silicon by fields induced by the gate array structure. After an accumulation time, which is short enough that most pixels contain no events and no pixels contain more than a single event, the charge is transferred across the detector to a readout array by manipulating the voltages on the gate array. Front-side illuminated CCDs have relatively low efficiency at lower energies as the X-ray must pass through the metallic gate array structure. Back-side illuminated devices, however, offer high quantum efficiency in the soft X-ray band (Fig. 54). CCDs also provide moderate energy resolution: FWHM $\sim$ 50 eV at 0.5 keV. Such spectral resolution enables analysis of the bright SWCX lines at energies greater than ~ 0.5 keV, O^{6+} (0.56 keV), O^{7+} (0.65 keV), Ne^{8+} (0.90 keV), Mg^{10+} (1.31 and 1.57 keV). At lower energies, however, the lines are strongly blended, even at much higher resolution. Spectroscopy at energies below ~ 0.3 keV is complicated by the incomplete collection feature which contains contributions from all energies as well as the particle background. Imaging at energies affected by the incomplete-collection feature thus has higher background and significantly lower contrast.

CCDs have three main disadvantages. First, they require cooling, typically to around 173 K or cooler. This temperature can be achieved with thermo-electric coolers (TECs) but requires radiators. Second, they suffer from radiation damage, which increases the charge transfer inefficiency (CTI) which, in turn, degrades the energy resolution, increases the minimum usable energy, and decreases the sensitivity to soft X-rays. Conventional X-ray telescopes minimize radiation damage by shielding the detector during radiation belt passages, the South Atlantic Anomaly (SAA), and other periods of increased energetic particle density. Shielding is accomplished either by rotating a filter wheel to a closed position or by moving the relatively small detector under a shield. For a wide-field imager the detector plane is large and has a large acceptance angle. As a result, implementing a moveable shield can be challenging. Third, CCDs are sensitive to optical light, which can be an issue for instruments which need to have look directions relatively close to the bright Earth.[1] Optical photons also create charge in the detector. If the accumulation time is too long, the charge accumulated due to optical photons can be a sizable fraction of that deposited by a soft X-ray. Applying a readout threshold is necessary to remove false detections, but also removes valid X-rays at the lowest energies.

APS sensors having single addressable pixels are intrinsically radiation hard compared with CCDs as there is no transfer of charge across the device. The pixels in an APS sensor can be read out much faster than a CCD which minimizes the difficulty with optical loading. The next generation of flagship X-ray observatories such as ESA's Athena mission, expected to launch in the late 2020s, is expected to use APS based detectors rather than CCDs. However, APS based detectors have not yet been flown and do not have a high TRL.

For soft X-rays, MCP detectors have lower intrinsic QE than back-side illuminated CCDs, though their QE generally increases towards lower energies. The last MCP to have been used in an astronomical X-ray observatory was the HRC on Chandra, which was a CsI coated detector and was not optimized for soft X-rays; it serves as our reference detector. KBr coatings can produce devices with significantly higher efficiency (Jelinsky et al. 1996). MCPs are large area detectors (up to 9×9 cm in the case of Chandra HRC-I) with precise photon localization, ~ 1 mm, which is still significantly smaller than the PSF of a wide-field X-ray optic. The time resolution for the detection of an event is in the microsecond range which allows MCPs to employ anti-coincidence shielding for background rejection. Compared with CCDs, MCPs are radiation hard, although they are not operable during radiation belt passages because their high voltage system needs to turned down (or off) to avoid damaging the detectors. MCPs do not need to be cooled.

MCP do have their disadvantages. They have very poor energy resolution; $\Delta E / E \sim 1$ at 1 keV, so there is no way of separating the desired charge exchange signal from the cosmic and other backgrounds spectroscopically. Further, while MCPs are insensitive to optical photons, they are very sensitive to UV and FUV photons, which can produce a substantial background when observing near the Earth.

For a potential near-future mission the advantages and disadvantages of an MCP detector versus CCD detector solution strongly depends on the available resources, spacecraft environment, and orbit, and will have to be studied on a case by case basis. CCDs may offer better performance in the short term but could be prone to faster degradation than MCPs and, hence, the total science return could be greater for an extended mission using MCPs.

7.2.3 Implementation

STORM As of the time of this writing, wide field-of-view micropore reflectors have flown in space on two rocket flights out of White Sands Missile Range. The first flight was the Sheath Transport Observer for the Redistribution of Mass (STORM) prototype that flew as a

---

[1] Similar problems arise when attempting to image Jupiter or Venus in the X-ray.

Fig. 55 DXL-2/CuPID flight data from the longitudinal scan across the helium focusing cone

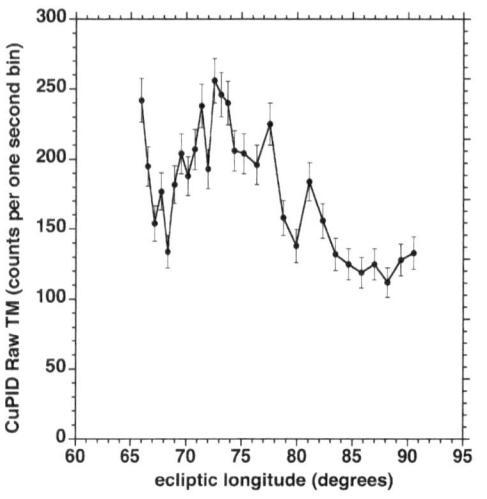

DXL_B_CuPID_raw_170823

piggy-back experiment on the Diffuse X-ray emission from the Local galaxy (DXL) mission on 12 December 2012. Although the DXL/STORM prototype was designed to accommodate nine 75-cm radius of curvature micropore reflectors for a total field-of-view of about $9° \times 9°$, budgetary constraints permitted population of only two facets for the flight. The DXL/STORM instrument viewed out the back of the rocket while the main DXL payload viewed out the side of the rocket to observe soft X-ray emission from the helium focusing cone. Consequently, DXL/STORM observed the soft X-ray background emission approximately perpendicular to the helium focusing cone during the flight. Collier et al. (2012) described DXL/STORM instrument development, while Collier et al. (2015) described the successful flight and analyzed the returned data. See Sects. 5.4 and 6.7 for more information about the helium focusing cone objectives of the DXL mission.

CuPID   The second rocket flight validating wide field-of-view soft X-ray imaging in space occurred when the Cusp Plasma Imaging Detector (CuPID), a 3U CubeSat form factor instrument, flew on the Diffuse X-ray emission from the Local galaxy-2 (DXL-2) mission on 4 December 2015. The CuPID instrument was not deployed from the rocket but rather made soft X-ray measurements in space while mounted to the rocket. This flight served as a validation in space of the Soft X-ray Instrument (SXI) imager to be flown on the free-flying 6U CuPID CubeSat (see below).

DXL-2/CuPID employed a single 55 cm radius of curvature facet providing about a $4.2° \times 4.2°$ field-of-view. Unlike DXL/STORM, DXL-2/CuPID viewed out the side of the rocket, in the same direction as the main payload. It therefore also observed in the direction of the helium focusing cone, allowing a comparison with observations from the main DXL-2 payload. Figure 55 shows the observed CuPID raw count rate from the DXL-2 longitudinal scan through the helium focusing cone. The nominal direction of the helium focusing cone based on a wide variety of different observations and observational techniques is from about 72° to about 79° ecliptic longitude (Witte et al. 1993; Witte 2004; Geiss and Witte 1996; Bertaux et al. 1985; Ajello et al. 1994; Möbius et al. 2012; Chalov 2014).

The Cusp Plasma Imaging Detector (CuPID) Cubesat Observatory is a 6U cubesat that will image the soft X-rays corresponding to ion density structures within the magnetospheric cusps, looking upward from a highly-inclined low Earth orbit platform. Ion dispersions have been observed in the cusp for several decades, but untangling their cause has remained impossible due to the space-time ambiguity with point measurements from individual spacecraft. With X-ray images this ambiguity is eliminated. The science goal of the CuPID mission is to distinguish the conditions under which magnetopause reconnection is patchy and bursty and the conditions under which reconnection is steady and continuous. As the mission will image the footprint of magnetopause reconnection, it will also monitor the growth or expansion of reconnection X-lines on the magnetopause. CuPID is led by Boston University and is under development for launch in 2019. The X-ray camera has a $4.6° \times 4.6°$ field of view and uses a single lobster-eye optical element to focus the X-rays.

SMILE   The Solar wind Magnetosphere Ionosphere Link Explorer (SMILE) is a joint European Space Agency (ESA) and Chinese Academy of Sciences (CAS) mission due for launch in late 2021 (Raab et al. 2016). SMILE is a small class mission, S2 in the ESA program, and has a science payload consisting of the UK-led Soft X-ray Imager (SXI), Canadian-led Ultraviolet Imager (UVI) and the Chinese-led Light Ion Analyser (LIA) and Magnetometer (MAG). Various orbits are being studied, but the basic parameters are highly elliptical (~ 19 R$_E$ apogee and ~ 1 R$_E$ perigee) with a high inclination ($> 69°$) above Earth's northern pole. The orbital period will be around 50 hours. The UVI will perform observations of the northern aurora simultaneously with SXI observations of the subsolar magnetopause and cusps. The two *in situ* devices, LIA and MAG, will provide contextual information.

The SXI is a compact design appropriate to the resource constraints of this mission class. The detector plane is formed by 2 CCDs, each with 4510×4510 native 18 μm pixels. In practice, the CCDs will operate with a factor of 6 on-chip binning and around one seventh of the CCD rows will be used as an asymmetric frame store. The CCDs are derivatives of the e2v CCD 270 devices being developed for the ESA M Class PLATO mission, but with structural changes to reduce the effect of Charge Transfer Inefficiency (CTI). These include adding a supplementary buried channel (SBC) in the image and store regions and using a narrow (20 μm) serial register. The CCDs have charge injection (CI) structures to further mitigate against CTI. The SXI has a radiation door mechanism with a thick aluminum flap (~ 10 mm) that will cover the CCDs at low altitudes (less than $\sim 50,000$ km) to protect the focal plane during radiation belt passages. Passive cooling will be used to reach the optimum operating temperature of -100 C. Figure 56 (left panel) shows a CAD model of the main instrument body.

The SXI mirror is an 8×4 MPO array that matches the oblong detector array. The MPOs have a focal length of 300 mm. The short focal length is necessary to achieve a usable FOV which is $26.5° \times 15.5°$ on the image plane. The FOV is orientated such that the short axis is parallel to the Earth-Sun line with the edge of the FOV $12.5°$ from the edge of the Earth. This configuration maximizes the observing efficiency of the instrument to the key targets (magnetosphere nose and cusps) given the practical constraints on the size and configuration of the instrument. This decision was strongly dependent on the details of the orbit, the pointing needs of the other instruments, and the key science goals.

The MPOs have a baseline configuration of 40 μm pores of length 1.2 mm, i.e., an L/D ratio of 30:1. This compromises some angular resolution for increased X-ray throughput compared with a configuration of L/D ~ 50:1 more usually adopted for astrophysics applications. The mirror total effective area is calculated to be ~ 14.8 cm^2 at 0.5 keV.

An additional feature of the SXI is the stray-light baffle and filter system. The baffle system is designed to minimize the amount of time that sunlight can directly enter the inside of the SXI telescope. When this becomes unavoidable due to the nature of the SMILE orbit and

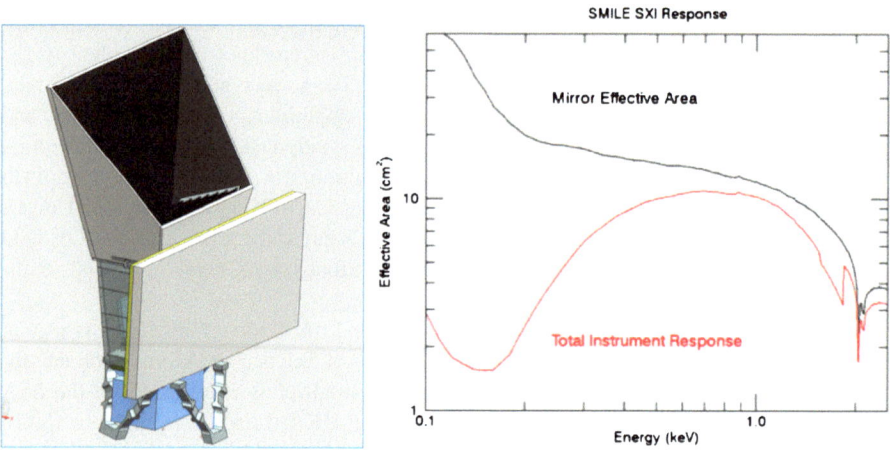

Fig. 56 *Left Panel*: CAD model of SXI showing baffle (top) main telescope tube (center) and front-end electronics box (bottom) with the radiator panel in front. The data processing unit and power supply box are not shown. *Right Panel*: Instrument response showing the mirror effective area (black) and total response including CCD QE and optical filter transmission (red)

pointing direction, the spacecraft can offset point to potential secondary science objectives. On the Earth-side the baffle is designed to intercept Earthlight with a low reflectively surface prior to reaching the optic plane. The baffle has interior planes orthogonal to the optic surface. Further Earthlight suppression is provided by the transmission properties of the MPO and a filter film that is baselined to be around 800 Å of aluminum. The total instrument response at the center of the FOV including the CCD QE, filter transmission and vignetting due to the stray-light baffle is shown in Fig. 56 (right panel).

7.3 Instrument Simulations

This section presents simulations of soft X-ray images for the predicted response of a base-line instrument similar to that shown in Fig. 53 (right panel). The base-line instrument includes a 10×10 facet nickel-plated MPO array with a radius of curvature of 100 cm, for a focal length of 50 cm. The resulting field of view is $20° \times 20°$, measuring the region with $< 25\%$ vignetting. From our tests of the instrument shown in the left panel of Fig. 53, the PSF FWHM is expected to be $\lesssim 15'$. The base-line detector is a 20 cm $\times$ 20 cm array of KBr coated MCP pairs. We have assumed that the response as a function of energy is similar to that of the Chandra HRI. The optical blocking filters are assumed to be similar to those flown on DXL/STORM and DXL/CuPID. These are composed of 2200 Å of polyimide and 300 Å of aluminum.

The first subsection describes how we simulate observations of the magnetosheath using a reasonable baseline instrument and reviews the features expected in the simulations. To reduce confusion, we will call an ideal X-ray image created from a MHD model a "model image", while that X-ray image after it has been convolved with the instrument responses, summed with the various background/foreground emission components, and sampled using the appropriate statistics will be called a "simulated image". The second subsection employs simulated images as a testbed for developing data analysis techniques to find the locations of the magnetopause and the bow shock.

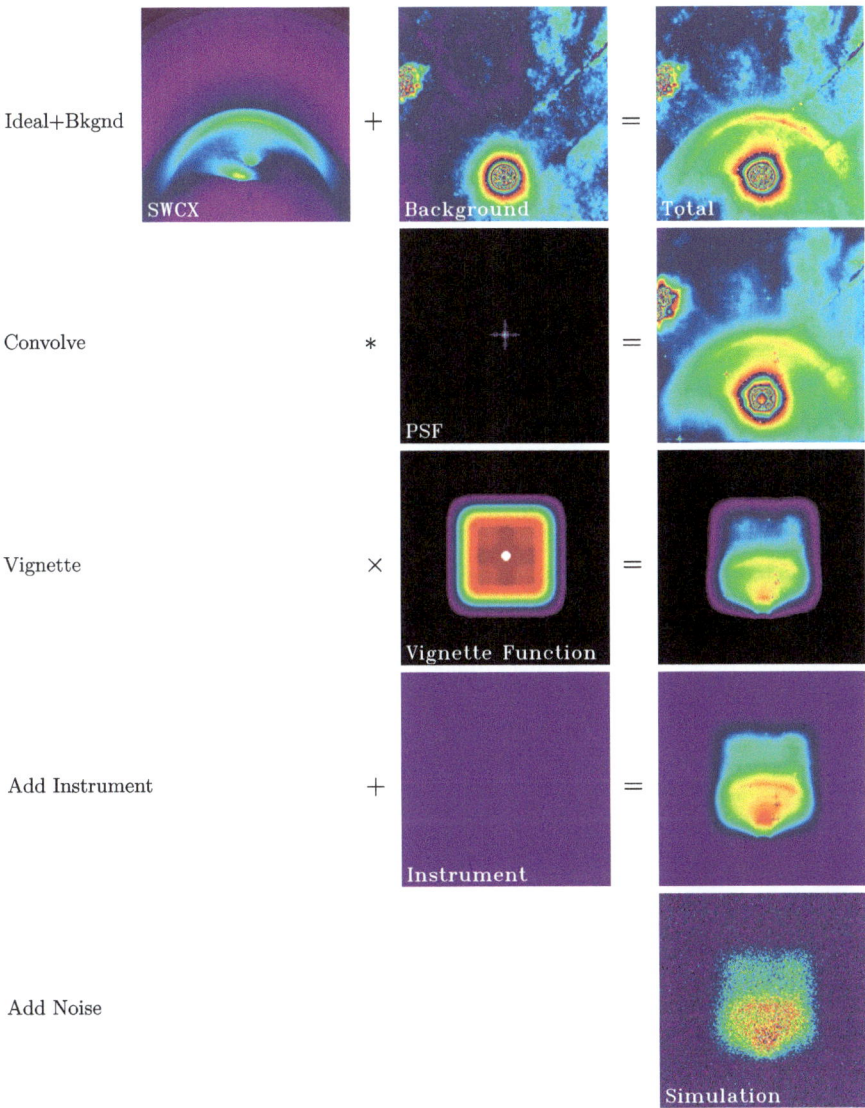

Fig. 57 A graphical demonstration of our method to construct a simulated image from the MHD based ideal X-ray image. All images have the same flux to color scaling. Integration times are two minutes. Colors run from near 0 (purple) to 30 (red) counts per $0.25° \times 0.25°$ pixel for the soft X-ray intensities and from 0 (black) to 1 (white) for the PSF and vignetting functions. Earth's brightness is not correctly represented because neither scattered solar X-rays nor time variable FUV lines are included in the figure

7.3.1 Creating the Simulations

Figure 57 schematically shows the process of going from an ideal X-ray image to a simulation of what a detector would register. The simulations use the solar wind proton density, speed, and pressure values from an MHD model cube produced by an MHD code. Each detector pixel corresponds to a line of sight through the model cube which is sampled at a resolution of 0.1 R_E or better. The neutral density is calculated from the Hodges (1973) model for the same locations, using the appropriate season and the solar f10.7 value. We then construct a mask to remove points within the inner boundary of the MHD model (typically $R_E < 2.5$), any region that would be occulted by the Earth, and regions of closed magnetic field lines where magnetospheric plasma populations are not expected to contain the high state ions required to produce X-ray emission. The initial model image, essentially a map of Q in units of $cm^{-4} s^{-1}$, is formed by summing $[n_p n_n v_{rel} \times mask]$ along each line of sight.

This Q map is converted directly to instrument counts for a given integration time using the conversion factor from Kuntz et al. (2015), which is to ROSAT count s^{-1} degree^{-2}. The conversion value was derived from ROSAT measurements of the magnetospheric SWCX emission from the flanks of the magnetosheath. One can convert from the ROSAT rate to any other particular instrument rate given a spectrum for the SWCX emission and the instrumental responses as a function of energy. Although the SWCX spectrum is not known extremely well, all that is needed for a reasonable conversion factor is the approximate spectral shape. Consistent with the linear correlation seen in Fig. 20 and results shown in Fig. 36 indicating that band-integrated emissions closely track solar wind proton fluxes for a wide range of fluxes and solar wind structures, we adopt a constant value for the band-integrated soft X-ray emissivity ς_n seen in Eq. (6).

X-rays from the magnetosheath are not the only ones that will be observed by a soft X-ray instrument. The cosmic X-ray background, both the diffuse component shown in Fig. 28 and the contribution from point sources shown in Fig. 30 will also be observed. Both were well measured by RASS at the energies of interest. Since the spectrum of the cosmic X-ray background is of great scientific interest, it is well known and conversion from ROSAT count rates to a particular instrument is trivial. It should be noted that the bulk of the heliospheric solar wind charge emission has not been removed from the ROSAT maps, so the typical contribution of the heliospheric SWCX to background is included. The near-Earth environment also emits strong line emissions in the far and extreme UV which can contaminate the X-ray band. We have used the neutral distribution from Hodges (1973) to calculate the Lyα emission, and IMAGE images to calculate the He$^+$ emission.

All of these components must be added to the X-ray emission from the magnetosheath. The sum of the photon components is then convolved with the point-spread function (PSF) and multiplied by the vignetting function (the change in the instrumental response with distance from the optical axis). To this image is then added the instrumental background, the signal due to energetic particles striking the detector directly *and* the signal due to X-rays that are produced when energetic particles strike material surrounding the detector.

This final model image, in count s^{-1} pixel^{-1} is then multiplied by the exposure time to determine the total number of counts in each pixel. We assume that the uncertainty in the number of counts in a pixel is given by the Poisson distribution. Using a random number generator one may extract from the model image a simulated image with the correct noise properties. Figure 57 graphically depicts the process of building a simulated image.

There are, however, some technical issues with this process. For example, most MHD codes do not track solar wind ions separately from other plasma population protons in the near-Earth environment. This lack of separation is problematic because the non-solar wind populations generally do not have the associated high state ions that are required for charge exchange. Thus, one must remove the non-solar wind plasma from the MHD output before calculating Q. For BATS-R-US, where the plasmasphere extends closer to the magnetopause than some other codes, we mask out the plasmasphere by removing the plasma on closed magnetic field lines, as determined by field line tracing. This does not work well around the cusps, but that discussion is beyond the scope of this section.

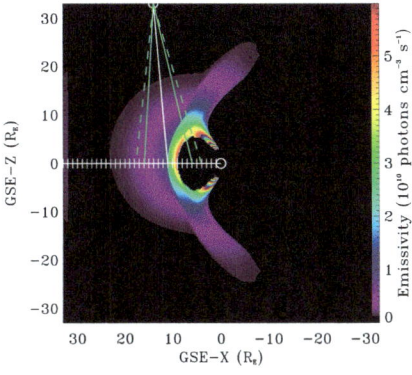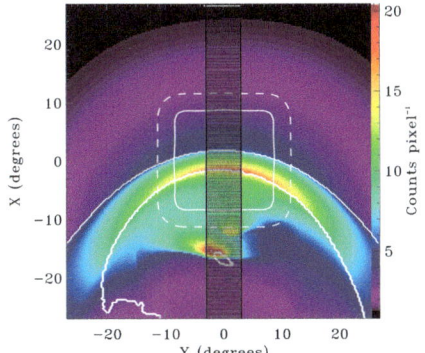

Fig. 58 Model results for $n_p = 12$ cm^{-3}, $v_p = 418$ km s^{-1} and $B_X, B_Y, B_Z = 0, 0, -5$ nT. The image was made with $0.25° \times 0.25°$ pixels. *Left Panel*: The X-ray emissivity in a noon-midnight slice through the magnetosheath. The location of the spacecraft and its field of view is shown. Note that the color scale wraps, producing stripes in the cusps. *Right Panel*: The magnetosheath as seen from the spacecraft at GSE $[X, Y, Z] = 36[\cos(67°), 0, \sin(67°)]$ R$_E$. This ideal X-ray image was made with $0.25° \times 0.25°$ pixels, no backgrounds, foregrounds, noise, vignetting, or PSF convolution. The *rounded white box* shows a typical instrumental field of view; the *solid line* denotes the region with an instrumental response within 90% of the peak response while the *dashed line* denotes the region with a response within 50% of the peak response. The *white curves* show the tangent to the magnetopause, as defined by the last closed field line, and the bow shock, as defined by the last element containing free-flowing solar wind. The *black boxes* show the locations of the pixels from which we extract the profile shown in Fig. 59

The left panel of Fig. 58 depicts the X-ray emissivity in a noon-midnight slice through the magnetosheath. The location of the observing spacecraft at GSE $(x, y, z) = 36[\cos(67°), 0, \sin(67°)]$ R$_E$ is also shown. The right panel of Fig. 58 shows what that spacecraft would see (without noise). White curves show the tangents to the magnetopause, as defined by the last closed field line, and the bow shock, as defined as the last grid element with the same velocity as the free-flowing solar wind. The dayside magnetosheath and cusps emit brightly. The solar wind region upstream from the subsolar bow shock and the flank magnetosheath produce relatively weak emissions. One can see that the X-ray structures, seen in projection, map very closely to the magnetopause and the bow shock. A wide range of Sun-Earth-space craft angles produce similar results.

The absolute location of the magnetopause can be inferred from X-ray observations by carefully considering the geometry. However, motion of the magnetopause can readily be inferred simply by following the time evolution of the X-rays.

The right panel of Fig. 58 shows a model of the X-ray emission as seen from a vantage point of GSE $(x, y, z) = 36[\cos(67°), 0, \sin(67°)]$ R$_E$. The left-hand panel of Fig. 59 shows a profile of the emission along the projected GSE-X axis. The dotted vertical lines show the location of the magnetopause (as determined from the tangent to the last closed field line), the peak of the emission, and the bow shock (as determined from the last element containing undeflected solar wind plasma). Note the secondary peaks due to the cusps at $(x, y) = (-12°, 2°)$ and $(-16°, -2°)$. In this case, the peak of the magnetosheath emission and the location of the magnetopause coincide. The bow shock corresponds to an inflection in the profile. From noiseless data one can easily determine the locations of the magnetopause and the bow shock using either the profile or its derivative.

The right-hand panel of Fig. 59 shows the profile extracted in the same way, observed from the same location, and in a model using the same solar wind conditions but a northward IMF orientation. We see that the peak in emissions is broader and that the line tangent to the

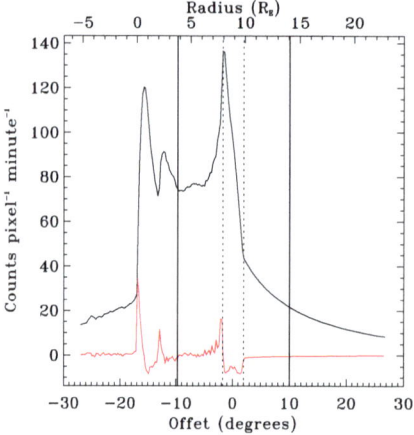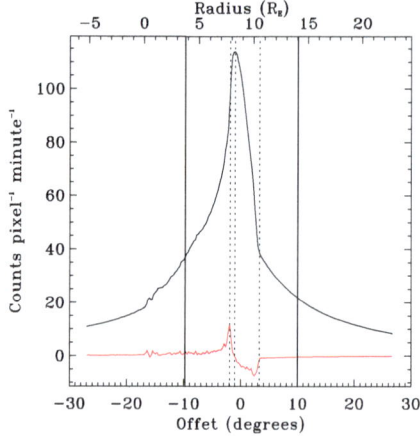

Fig. 59 The X-ray emission profile along the projected GSE-X axis. *Left Panel*: Profile for the 6° wide region shown in Fig. 58. *Black curve*: the model profile. *Red Curve*: the derivative of the model profile. *Solid black vertical lines*: the limit of the region for which the vignetting function is > 0.9. The *Dashed black vertical lines*: the model locations of the magnetopause (offset = −1.88°), the peak of the emission (offset = −1.75°, note that this line overlays the line for the magnetopause), and the bow shock (offset = 2.0°). The IMF is southwards in this model. *Right Panel*: the profile taken from the same location for a model with the same solar wind conditions but with a northward IMF. In this case the model locations are at offsets of −1.88° for the magnetopause, −1.0° for the peak of the emission, and 3.25° for the bow shock

magnetopause lies slightly earthward from the peak of the emission. This is consistent with the formation of a depletion layer as expected for northward IMF orientations where the offset increases as the solar wind plasma flux decreases. For the profiles produced by BATS-R-US, the magnetopause coincides with a strong positive gradient in emissions located just inside the peak location, but the behavior may depend upon the MHD model used. Note as well the lack of significant cusp emissions.

For strongly northward IMF, finding the magnetopause location, in the absolute sense, becomes more difficult, because locating a sharp drop in the presence of noise is difficult. However, tracking the relative location of the magnetopause by determining the peak of the emission is still relatively easy to do, particularly at stronger solar wind fluxes.

7.3.2 Data Analysis with Simulations

Starting with the images generated by the procedure illustrated in Fig. 57, we examine how well physical parameters can be extracted from simulated observations. The model image is in counts, and the uncertainty is assumed to be governed by counting statistics. Each pixel in the simulated image is drawn from the Poisson distribution whose mean is given by the value of that pixel in the model image.

The first step in the analysis is to remove the backgrounds. The instrumental background presumably can be measured on orbit. The cosmic X-ray background has already been measured, though in a slightly different band-pass, by ROSAT, and does not change with time. The FUV background has been measured by IMAGE and changes with solar wind conditions. Unless simultaneous FUV observations are available from the same vantage point, the FUV background will likely be the most difficult to remove. For the data analysis, backgrounds were treated in the same way as model images. That is, the photon backgrounds were convolved with the PSF, multiplied by the vignetting function, and added to the flat

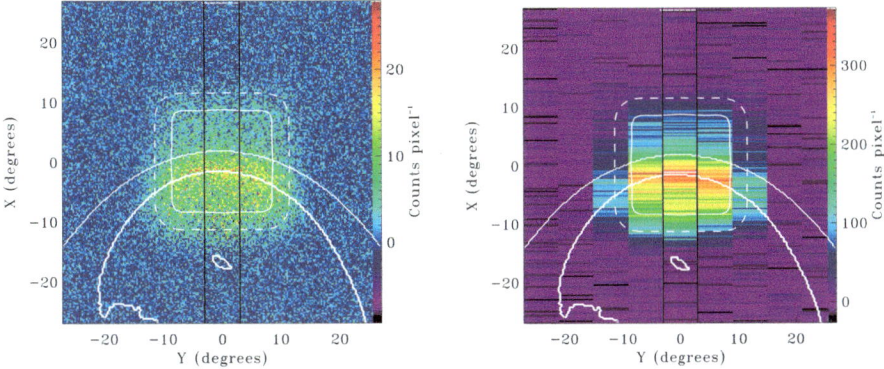

Fig. 60 *Left Panel*: Simulated image drawn from the model image from Fig. 57 with the addition of backgrounds, convolution with the point spread function, and multiplication by the vignetting function, and the addition of instrumental noise. *Right Panel*: the simulated image rebinned to preserve the high resolution perpendicular to the magnetopause but maintain high signal-to-noise values in each pixel. The binned pixels are $6° \times 0.25°$

instrumental background. This background image is then subtracted from the simulated image.

Since the physical pixels of the detectors are smaller than the resolution, the simulated image can be rebinned to preserve the best possible resolution perpendicular to the magnetopause while increasing the signal to noise in the profile across the magnetosheath. This rebinning requires pixels to have a width comparable to the resolution, and a length short enough not to mix emission from different layers of the magnetosheath. In practice, a pixel that is $0.25°$ wide in the direction normal to the magnetopause, and $4°$ to $6°$ long in the direction along the magnetopause works well. For a spacecraft 33 R_E from the nose of the magnetosheath, this angular size translates to linear dimensions of $0.144\ R_E$ by $3.45\ R_E$ at the magnetopause. Since the rebinning is done in software, one could imagine even longer pixels that are deformed to follow the curvature of the magnetopause. The curvature would have to be determined from the data and might require computationally intensive optimization routines. However, such long pixels might obscure asymmetries or small scale features ($\sim 3\ R_E$) on the magnetopause.

Clearly, determining the location of the magnetopause, the peak of the emission, and the bow shock in the presence of noise is not possible using the derivative. As can be seen from Figs. 60 and 61, the separation between the magnetopause and the peak of the emission is quite small. Thus we will concentrate on the level of precision with which one can determine the location of the peak of the emission, and then consider the implications for the magnetopause and the bow shock.

One can imagine a number of possible algorithms for determining the location of the peak emission. The location of the brightest pixel is a very poor measure of the peak even at relatively high count rates. Taking the maximum pixel of a smoothed profile is an improvement, but still has a relatively large uncertainty. Because the peak is asymmetric, determination of the peak through a weighted mean or fitting a Gaussian is ill-advised, as both methods assume a symmetric peak. Cross-correlation of the measured profile and a model profile can find the peak very precisely if one already knows the model profile, or can create an appropriate guess.

As illustrated in Fig. 62, true cross-correlation takes the product of two series, our profiles, as a function of the relative shift between the two. Due to the vignetting function

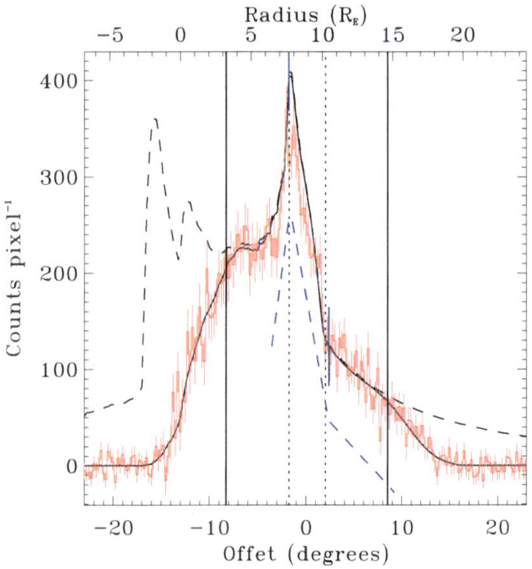

Fig. 61 The profile from the central column of pixels. *Black dashed curve*: model profile *Black solid curve*: model profile after vignetting *Red curve with error bars*: simulated profile with Poisson uncertainties *Blue dashed curve*: "broken linear fit" to the peak and the bow shock offset downwards for clarity. *Black solid verticals*: the limits of the region where the vignetting function > 0.9. *Black dashed verticals*: the model locations of the peak and the bow shock. *Blue solid verticals*: measured locations of peak and bow shock

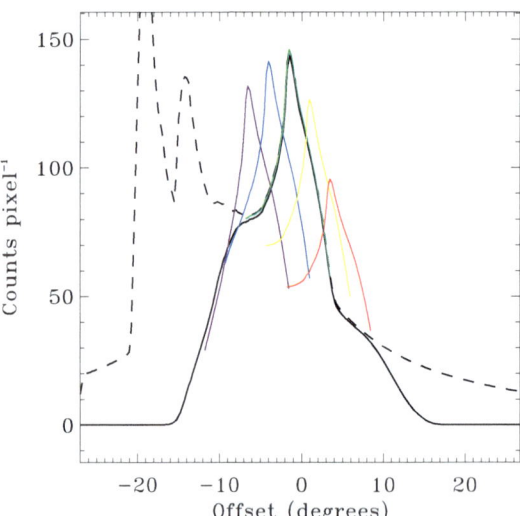

Fig. 62 A graphical demonstration of the cross-correlation technique for finding the peak. For clarity, we have shown the process without noise. The dashed line is the true profile while the solid black line is the profile after vignetting, the equivalent to the measured profile. The colored profiles are the center of the model profile shifted and, at each shift position, multiplied by the vignetting function and scaled to best match the measured profile. In the absence of significant noise, the best χ^2 as a function of shift is a very sharply peaked function

and the presence of the cusps, implementing the cross-correlation is not so simple. We first remove the parts of the measured profile dominated by the cusps or where the vignetting function is < 0.75. The model profile with which we wish to compare must be rather narrow to avoid windowing effects; we generally use a model profile that encompasses ±5° around the peak in the X direction along the Sun-Earth line. For these simulations, the shape of the model profile has been chosen to be the same as that of the profile from which the simulated data were created. While this assumes knowledge that we would not have with real data this issue is discussed below. For each relative shift between the measured and model profiles, one must apply to the model the same vignetting function experienced by the measured profile. Thus, the shape of the model profile changes with the relative shift. Finally, the model profile must be scaled to the measured profile, and that scaling will also change with the

 ⚛ Springer

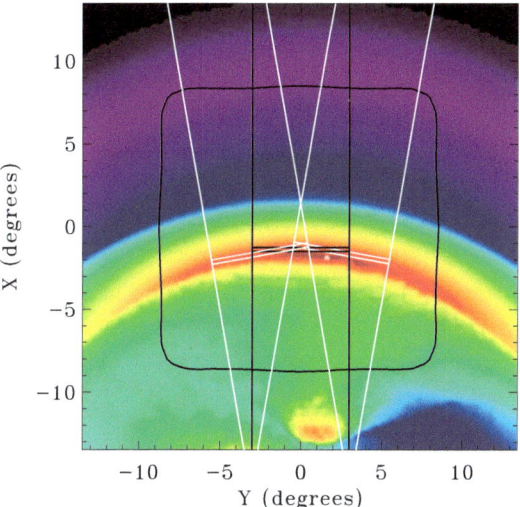

Fig. 63 Demonstration of sampling the front over a narrow range of angles, which improves the accuracy of the peak emission

relative shift. Thus, while we are doing the equivalent of a cross-correlation, we are actually fitting the model profile to the measured profile for each relative shift, and determining the peak from the best fit.

There are a number of caveats. First, even with a large number of counts, the FOV may not be broad enough to capture the bulk of the magnetosheath emission profile, in which case no algorithm will produce a reasonable measure. The vignetting function should be flat for roughly twice the width of the magnetosheath or else the shape of the peak and the shape of the vignetting function become confused. On the Earthward side, the profile should extend to at least the minimum between the cusps and the magnetosheath; on the Sunward side, the profile should extend at least to the bow shock, if one is interested only in the location of the magnetopause, and $\sim 10°$ beyond the bow shock if one is interested in determining the location of the bow shock.

No matter what method is used to determine the peak of the profile, one can increase the accuracy by measuring the location of the peak along multiple cuts. Figure 63 shows an example of such a multiple measurement; over a model image we have set down three strips from which to extract profiles. In this case, the strips are pivoted on the location of the Earth and are rotated $\pm 10°$ from the strip through the magnetosheath. The location of the peak can be determined for each of these strips, and those values combined in a weighted sum to determine the location of the peak emission. The configuration shown decreases the uncertainty in the location of the peak by a factor of ~ 1.5. Ideally, one would want to measure the curvature of the front and measure any deformations, so this technique must be applied cautiously.

Figure 64 shows the dependence of the uncertainty in determining the peak location as a function of the number of counts in the pixel at the peak emission. Each point represents the dispersion of the measurement of the peak location from 1000 simulations of the same model image. The boxes represent the uncertainty in the peak position as derived from a trio of cross-correlation measurements near the nose of the magnetopause. The crosses are from single cross-correlation measurements and are a factor of ~ 1.5 larger. Between solar wind fluxes of $\sim 1.8 \times 10^8$ cm^{-2} s^{-1} and $\sim 15 \times 10^8$ cm^{-2} s^{-1}, the uncertainty at a particular exposure time declines as roughly 1/peak counts. At lower solar wind fluxes, the uncertainty is worse than the linear relation, probably because the magnetosheath emission is comparable

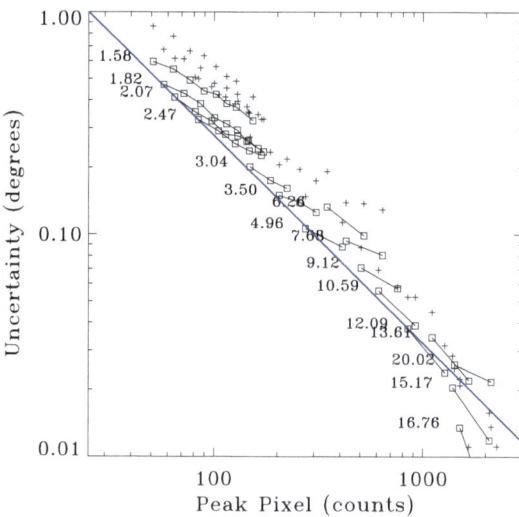

Fig. 64 The accuracy with which one can determine the location of the peak emission as a function of the true peak count rate. The different symbols represent different methods of peak location; *Crosses*: cross-correlation. *Squares*: cross-correlation employing three cuts through the magnetopause, averaged. This plot was assembled from simulations at different solar wind fluxes (the numbers in the plot multiplied by 10^8) and exposure times, typically 2, 2.5, 3, 3.5, 4, 4.5, 5, and 6 minutes. For the squares, the lines connect simulations with the same flux but different exposure times. The blue line is an arbitrary linear function in log-log space that roughly describes the lower envelope of the data points

to the background at these solar wind fluxes. Note that increasing the exposure time is not as effective at reducing the uncertainty as is increasing the solar wind flux, so exposure time and solar wind flux are not fungible. This effect is possibly due to the magnetosheath becoming narrower at higher fluxes. The profile was well centered in the FOV for all of these simulations, so vignetting was not an issue.

As can be seen from Fig. 64, uncertainties in determining the location of the peak emission range from $\sim 1°$ for very low count rates to 0.01 degrees for very high count rates. Uncertainties for two-minute exposures are about 0.4° to 0.5° for nominal solar wind fluxes of 2.0×10^8 cm^{-2} s^{-1}, but only $\sim 0.1°$ for the 5.0×10^8 cm^{-2} s^{-1} flux used in constructing Figs. 58–63. From a distance of 30 R$_E$, uncertainties of 0.4° and 0.1° in angular resolution correspond to uncertainties of 0.2 to 0.05 R$_E$ in the location of the magnetopause, much less than the ~ 1 R$_E$ nominally associated with magnetic erosion events, and therefore sufficient to conduct heliophysics research.

Note that one must have a reasonably correct template for the cross correlation. For the baseline tests in Fig. 64 we used the true profile, which assumes knowledge that we will not have with real data. To understand the extent of the problem, we used profiles from models with different solar wind conditions. We found that for lower solar wind fluxes ($nv < 4 \times 10^8$ cm^{-2} s^{-1}) an incorrect choice of template increased the uncertainty by no more than a factor of ~ 1.3. In general, the greater the disparity between the true profile and the model profile, the larger the increase in the uncertainty, the larger the systematic offsets between the true and measured locations, and the poorer the minimization. These differences become larger at higher count rates and exposures. Thus, it is important to have the correct template at higher solar wind fluxes, but not so important at lower solar wind fluxes.

Fig. 65 The accuracy with which one can determine the location of the bow shock as a function of the counts in the peak pixel. *Crosses*: the bow shock location was determined by the broken line fitting method. *Boxes*: the bow shock location as determined from the mean of three measurements through the subsolar magnetosheath. The blue line is an arbitrary scaling of 1/peak counts. Numbers in the figure indicate solar wind fluxes ($\times 10^8$). Line segments connect simulation results with similar fluxes but different integration times

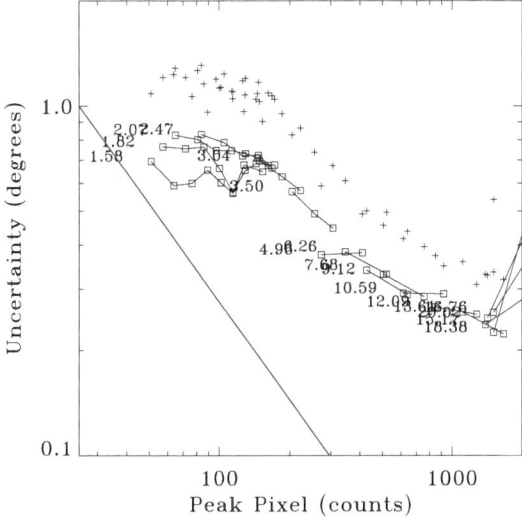

Choosing the correct templates is not always straightforward. Different MHD codes produce somewhat different profiles, as do different models for the neutral distribution. Determining which is correct is a matter that must be addressed observationally. The longer a magnetosheath imager operates, the larger the library of observed profiles, and the higher the precision in measuring the location of the peak emission. It will be necessary to generate a library of profiles from the coaddition of images for similar solar wind conditions (probably nv and $\mathbf{B}/|\mathbf{B}|$). However, if the solar wind flux is not changing rapidly, the application of a single template to a series of images, *even if the template is incorrect*, should correctly extract the motion of the magnetopause.

With sufficient counts, one can also determine the location of the bow shock. As can be seen in Fig. 59, the bow shock coincides with an inflection in the emission profile. In the derivative of the profile, this inflection becomes even sharper. Since the derivative of the profile changes very quickly just inside the bow shock, one can locate the inflection effectively by searching for the maximum radius for which the derivative is $0.2\times$ the maximum derivative in the radial profile. There are likely methods to locate the inflection more precisely, but the current method is quite robust. Locating the inflection in real data is more difficult. A rough solution can be found by fitting a straight line with a break, or hinge, to the profile at offset angles/radii greater than the peak. (The result of applying this "broken linear fit" technique can be seen in Fig. 61) When the location of the hinge is one of the fit parameters, the fitted location tracks the offset angle/radius at which the bow shock occurs (as determined from the derivative) reasonably well. The accuracy as a function of the peak counts is shown in Fig. 65; it declines more slowly than 1/peak counts. Uncertainties in determining the bow shock position for two-minute integration times range from $\sim 0.8°$ ($0.4\,\mathrm{R}_E$) for a solar wind flux of $2.0 \times 10^8\,\mathrm{cm}^{-2}\,\mathrm{s}^{-1}$ to $\sim 0.4°$ ($0.2\,\mathrm{R}_E$) for the $5.0 \times 10^8\,\mathrm{cm}^{-2}\,\mathrm{s}^{-1}$ solar wind flux used to construct Figs. 58 through 63. A cross-correlation method would probably work better, but only if the instrument FOV extends to sufficiently high radii. Fitting a more complex parameterization of the profile may work better as it might not require as great an extent of the lever-arm at high offset angles/radii.

Intervals in which the solar wind flux remains steady while the IMF orientation changes abruptly, like that reported by Aubry et al. (1970), will be of great use in identifying the effects of magnetic reconnection upon the dayside magnetopause. To determine how often

Fig. 66 The fraction of events for which $\Delta R_{MP} < 0.2$ over a time interval $> T$ as a function of R_{MP}. (For $R_{MP} = 9$ R$_E$, 60% of events have R_{MP} quasi-stationary for longer than 2.5 minutes.) The red line shows the fraction of time that R_{MP} is less than a given value

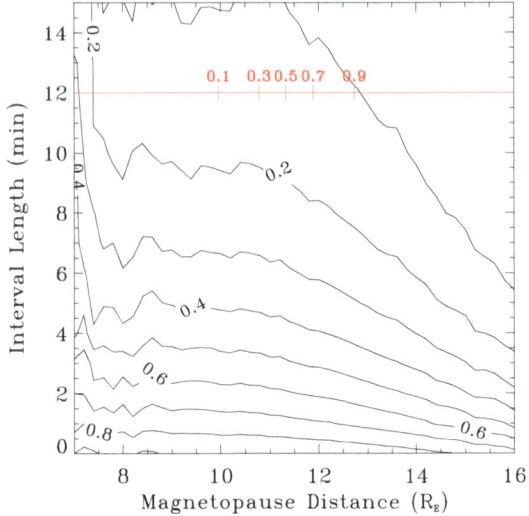

such intervals occur and evaluate the relationship between the uncertainty in the measurement of the peak location and the combination of solar wind flux and exposure time, we must determine the typical time scales over which the magnetopause remains quasi-stationary. We used all the 1-minute OMNI data from January 1995 to Dec. 2016 to calculate the magnetopause standoff distance, R_{MP}, for every point with a valid proton density and velocity by scaling locations to the sixth root of the solar wind dynamic pressure. For each point we determined the Δt over which R_{MP} changed by less than ± 0.1 R$_E$. We then created the distribution of Δt as a function of R_{MP}, which we then plotted as the fraction of events with $\Delta t > T$ as a function of R_{MP} (Fig. 66) We predict that for the most likely values of R_{MP}, that R_{MP} is quasi-stationary for greater than one minute $\sim 75\%$ of the time, greater than two minutes $\sim 65\%$ of the time, and greater than 3 minutes $\gtrsim 50\%$ of the time. Quasi-stationary intervals of 6–8 minutes are not uncommon. We conclude that it will frequently be possible to examine the effect of one solar wind parameter upon boundary locations while keeping others nearly constant.

Summarizing, this section has demonstrated that there will be clearly identifiable structures in soft X-ray images that correspond to the bow shock and magnetopause. Determining the precise locations of these boundaries will require comparisons with the profiles from global numerical simulations and/or statistical studies of the structures expected as a function solar wind conditions. Tracking the motion of these structures and determining changes in the shape of the bow shock and magnetopause as a function of time will be far easier. A comparison of simulation results with observations of solar wind variability indicates that the necessary measurements can be made on the time and spatial scales appropriate for magnetospheric studies.

8 Summary and Prospects

Both the Heliophysics and Planetary Physics disciplines seek to understand the nature of the solar wind's interaction with solar system obstacles such as Earth's magnetosphere, the Venusian and Martian ionospheres, and comets. Understanding these interactions is important not only because it can help determine the space weather environment and rate of atmospheric loss at each obstacle, but also because it will provide information concerning fundamental plasma physics processes such as magnetic reconnection and particle energization. Each of the many interaction processes proposed to occur at these obstacles creates a host of plasma structures, including bow shocks, magnetopauses, ionopauses, and cusps. Although *in situ* measurements confirm the occurrence of the various interaction mechanisms, it can be difficult to infer the extent and therefore importance of each interaction mode from localized measurements.

Global measurements of the relevant plasma structures offer a more direct means for determining the nature of the solar wind-obstacle interactions. Numerical simulations that employ magnetohydrodynamic models for solar wind densities and velocities and the results from a Monte Carlo simulation of exospheric densities indicate that the plasma density structures can be imaged in the soft (0.1–2.0 keV) X-rays generated when high charge state solar wind ions exchange electrons with exospheric neutrals. Theory indicates that the emissions should always be present, at wavelengths and intensities that depend upon the composition and density of high charge state solar wind ions. Past observations by narrow field-of-view soft X-ray telescopes on astrophysics missions demonstrate that Earth's cusps, comets, and the magnetosheaths of Venus, Earth, and Mars, emit soft X-rays at the expected wavelengths and intensities that can greatly exceed those from the soft X-ray background. Despite varying solar wind compositions, the same observations indicate that the bandwidth-integrated intensities closely track the solar wind density. Finally, the simulations indicate that features corresponding to the bow shock, magnetopause, and edges of the cusps should be readily identifiable in soft X-ray images of the dayside interaction, particularly during the most interesting intervals when solar wind densities are large and conditions are most dynamic.

The recent development of wide FOV soft X-ray telescopes offers a unique opportunity to observe the plasma structures generated by the solar wind's interaction with heliospheric obstacles. Simulation results reported here and elsewhere demonstrate that instruments with lobster-eye optics that focus grazing incident X-rays onto position-sensitive detector planes can image the structures with the cadences and spatial resolutions needed to track bow shock, magnetopause, and cusp motion. Wide FOV soft X-ray imagers were recently selected to fly on the forthcoming Cusp Plasma Imaging Detector (CuPID) Cubesat observatory and Solar wind Magnetosphere Ionosphere Link Explorer (SMILE Branduardi-Raymont and Wang 2016) missions in 2019 and 2021, respectively. CuPID will determine the nature of reconnection at Earth's magnetopause by imaging structures within the Earth's cusps from below while moving along a low-altitude Sun-synchronous orbit. SMILE will image the subsolar bow shock, magnetosheath, and magnetopause, as well as the cusps from a high-altitude high-inclination orbit that takes the spacecraft outside the bow shock. As summarized in this paper, there are strong scientific reasons that would justify future soft X-ray missions to Venus, Mars, and comets. Tomography would reveal the details of the plasma structures that diagnose the interactions at each of these obstacles and the Earth.

Acknowledgements   First and foremost we wish to thank ISSI for their kind hospitality for the two meetings over which the bulk of this publication was crafted. We would also like to thank the New Deal Café of Greenbelt Maryland, on whose premises many fruitful scientific discussions were held. MRC acknowledges William Farrell, Timothy Stubbs, and Menelaos Sarantos for their helpful comments. Research at GSFC was funded by NASA GSFC's IRAD, STG, and SIF programs and the XMM-Newton program.

References

M.H. Acuna, J.E.P. Connerney, P. Wasilewski, R.P. Lin, K.A. Anderson, C.W. Carlson, J. McFadden, D.W. Curtis, D. Mitchell, H. Reme, C. Mazelle, J.A. Sauvaud, C. D'Uston, A. Cros, J.L. Medale, S.J. Bauer, P. Cloutier, M. Mayhew, D. Winterhalter, N.F. Ness, Magnetic field and plasma observations at Mars: initial results of the Mars Global Surveyor mission. Science **279**, 1676 (1998). https://doi.org/10.1126/science.279.5357.1676

J.M. Ajello, W.R. Pryor, C.A. Barth, C.W. Hord, A.I.F. Stewart, K.E. Simmons, D.T. Hall, Observations of interplanetary Lyman-alpha with the Galileo Ultraviolet Spectrometer: multiple scattering effects at solar maximum. Astron. Astrophys. **289**, 283–303 (1994)

A.L. Albee, R.E. Arvidson, F. Palluconi, T. Thorpe, Overview of the Mars Global Surveyor mission. J. Geophys. Res. **106**, 23291–23316 (2001). https://doi.org/10.1029/2000JE001306

C.J. Alexander, C.T. Russell, Solar cycle dependence of the location of the Venus bow shock. Geophys. Res. Lett. **12**, 369–371 (1985). https://doi.org/10.1029/GL012i006p00369

C.J. Alexander, J.G. Luhmann, C.T. Russell, Interplanetary field control of the location of the Venus bow shock—evidence for comet-like ion pickup. Geophys. Res. Lett. **13**, 917–920 (1986). https://doi.org/10.1029/GL013i009p00917

I.I. Alexeev, D.G. Sibeck, S.Y. Bobrovnikov, Concerning the location of magnetopause merging as a function of the magnetopause current strength. J. Geophys. Res. **103**, 6675–6684 (1998). https://doi.org/10.1029/97JA02863

R.C. Allen, S.A. Livi, J. Goldstein, Variations of oxygen charge state abundances in the global magnetosphere, as observed by Polar. J. Geophys. Res. Space Phys. **121**, 1091–1113 (2016a). https://doi.org/10.1002/2015JA021765

R.C. Allen, S.A. Livi, S.K. Vines, J. Goldstein, Magnetic latitude dependence of oxygen charge states in the global magnetosphere: insights into solar wind-originating ion injection. J. Geophys. Res. Space Phys. **121**, 9888–9912 (2016b). https://doi.org/10.1002/2016JA022925

R.C. Allen, S.A. Livi, S.K. Vines, J. Goldstein, I. Cohen, S.A. Fuselier, B.H. Mauk, H.E. Spence, Storm-time empirical model of O^+ and O^{6+} distributions in the magnetosphere. J. Geophys. Res. **122**, 8353–8374 (2017). https://doi.org/10.1002/2017JA024245

E. Amata, V. Formisano, R. Cerulli-Irelli, P. Torrente, A.D. Johnstone, A. Coates, B. Wilken, K. Jockers, J.D. Winningham, D. Bryant, The cometopause region at Comet Halley, in *ESLAB Symposium on the Exploration of Halley's Comet*, ed. by B. Battrick, E.J. Rolfe, R. Reinhard. ESA Special Publication, vol. 250 (1986)

U.V. Amerstorfer, N.V. Erkaev, D. Langmayr, H.K. Biernat, On Kelvin Helmholtz instability due to the solar wind interaction with unmagnetized planets. Planet. Space Sci. **55**, 1811–1816 (2007). https://doi.org/10.1016/j.pss.2007.01.015

U.V. Amerstorfer, N.V. Erkaev, U. Taubenschuss, H.K. Biernat, Influence of a density increase on the evolution of the Kelvin-Helmholtz instability and vortices. Phys. Plasmas **17**(7), 072901 (2010). https://doi.org/10.1063/1.3453705

B.J. Anderson, S.A. Fuselier, Magnetic pulsations from 0.1 to 4.0 Hz and associated plasma properties in the Earth's subsolar magnetosheath and plasma depletion layer. J. Geophys. Res. **98**, 1461–1479 (1993). https://doi.org/10.1029/92JA02197

B.J. Anderson, T.-D. Phan, S.A. Fuselier, Relationships between plasma depletion and subsolar reconnection. J. Geophys. Res. **102**, 9531–9542 (1997). https://doi.org/10.1029/97JA00173

K.A. Anderson, J.H. Binsack, D.H. Fairfield, Hydromagnetic disturbances of 3- to 15-minute period on the magnetopause and their relation to bow shock spikes. J. Geophys. Res. **73**, 2371–2386 (1968). https://doi.org/10.1029/JA073i007p02371

J.R.P. Angel, Lobster eyes as X-ray telescopes. Astrophys. J. **233**, 364–373 (1979). https://doi.org/10.1086/157397

C.D. Anger, S.K. Babey, L.L. Cogger, A.L. Broadfoot, R.G. Brown, An ultraviolet auroral imager for the Viking spacecraft. Geophys. Res. Lett. **14**, 387–390 (1987). https://doi.org/10.1029/GL014i004p00387

M.P. Aubry, C.T. Russell, M.G. Kivelson, Inward motion of the magnetopause before a substorm. J. Geophys. Res. **75**, 7018 (1970). https://doi.org/10.1029/JA075i034p07018

L.A. Avanov, S.A. Fuselier, O.L. Vaisberg, High-latitude magnetic reconnection in sub-Alfvénic flow: interball tail observations on May 29, 1996. J. Geophys. Res. **106**, 29491–29502 (2001). https://doi.org/10.1029/2000JA000460

J. Bailey, M. Gruntman, Experimental study of exospheric hydrogen atom distributions by Lyman-alpha detectors on the TWINS mission. J. Geophys. Res. Space Phys. **116**, 9302 (2011). https://doi.org/10.1029/2011JA016531

H. Balsiger, K. Altwegg, F. Buhler, J. Geiss, A.G. Ghielmetti, B.E. Goldstein, R. Goldstein, W.T. Huntress, W.-H. Ip, A.J. Lazarus, A. Meier, M. Neugebauer, U. Rettenmund, H. Rosenbauer, R. Schwenn, R.D.

Sharp, E.G. Shelly, E. Ungstrup, D.T. Young, Ion composition and dynamics at comet Halley. Nature **321**, 330–334 (1986). https://doi.org/10.1038/321330a0

S.J. Bame, J.R. Asbridge, W.C. Feldman, E.E. Fenimore, J.T. Gosling, Solar wind heavy ions from flare-heated coronal plasma. Sol. Phys. **62**, 179–201 (1979). https://doi.org/10.1007/BF00150143

W. Baumjohann, R.A. Treumann, *Basic Space Plasma Physics* (1996)

M.W. Bautz, E.D. Miller, J.S. Sanders, K.A. Arnaud, R.F. Mushotzky, F.S. Porter, K. Hayashida, J.P. Henry, J.P. Hughes, M. Kawaharada, K. Makashima, M. Sato, T. Tamura, Suzaku observations of Abell 1795: cluster emission to r_{200}. Publ. Astron. Soc. Jpn. **61**, 1117 (2009)

P. Beiersdorfer, J.K. Lepson, G.V. Brown, S.B. Utter, S.M. Kahn, D.A. Liedahl, C.W. Mauche, Observation of quasi-continuum line emission from Fe VII to Fe X in the extreme-ultraviolet region below 140 Å. Astrophys. J. Lett. **519**, 185–188 (1999). https://doi.org/10.1086/312122

P. Beiersdorfer, G.V. Brown, J.J. Drake, M.-F. Gu, S.M. Kahn, J.K. Lepson, D.A. Liedahl, C.W. Mauche, D.W. Savin, S.B. Utter, B.J. Wargelin, Emission line spectra from low-density laboratory plasmas, in *Revista Mexicana de Astronomia y Astrofisica Conference Series*, ed. by S.J. Arthur, N.S. Brickhouse, J. Franco. Revista Mexicana de Astronomia y Astrofisica Conference Series, vol. 9 (2000), pp. 123–130

P. Beiersdorfer, C.M. Lisse, R.E. Olson, G.V. Brown, H. Chen, X-ray velocimetry of solar wind ion impact on comets. Astrophys. J. Lett. **549**, 147–150 (2001). https://doi.org/10.1086/319143

P. Beiersdorfer, K.R. Boyce, G.V. Brown, H. Chen, S.M. Kahn, R.L. Kelley, M. May, R.E. Olson, F.S. Porter, C.K. Stahle, W.A. Tillotson, Laboratory simulation of charge exchange-produced X-ray emission from comets. Science **300**, 1558–1560 (2003). https://doi.org/10.1126/science.1084373

M. Benna, D.M. Hurley, T.J. Stubbs, P.R. Mahaffy, R.C. Elphic, Observations of meteoroidal water in the lunar exosphere by the LADEE NMS instrument, in *Annual Meeting of the Lunar Exploration Analysis Group*. LPI Contributions, vol. 1863 (2015a), p. 2059

M. Benna, P.R. Mahaffy, J.S. Halekas, R.C. Elphic, G.T. Delory, Variability of helium, neon, and argon in the lunar exosphere as observed by the LADEE NMS instrument. Geophys. Res. Lett. **42**, 3723–3729 (2015b). https://doi.org/10.1002/2015GL064120

J. Berchem, C.T. Russell, Flux transfer events on the magnetopause—spatial distribution and controlling factors. J. Geophys. Res. **89**, 6689–6703 (1984). https://doi.org/10.1029/JA089iA08p06689

J. Berchem, R.L. Richard, C.P. Escoubet, S. Wing, F. Pitout, Asymmetrical response of dayside ion precipitation to a large rotation of the IMF. J. Geophys. Res. Space Phys. **121**, 263–273 (2016). https://doi.org/10.1002/2015JA021969

J.L. Bertaux, R. Lallement, V.G. Kurt, E.N. Mironova, Characteristics of the local interstellar hydrogen determined from PROGNOZ 5 and 6 interplanetary Lyman-alpha line profile measurements with a hydrogen absorption cell. Astron. Astrophys. **150**, 1–20 (1985)

C. Bertucci, C. Mazelle, M. Acuña, Structure and variability of the Martian magnetic pileup boundary and bow shock from MGS MAG/ER observations. Adv. Space Res. **36**, 2066–2076 (2005). https://doi.org/10.1016/j.asr.2005.05.096

C. Bertucci, F. Duru, N. Edberg, M. Fraenz, C. Martinecz, K. Szego, O. Vaisberg, The induced magnetospheres of Mars, Venus, and Titan. Space Sci. Rev. **162**, 113–171 (2011). https://doi.org/10.1007/s11214-011-9845-1

G.L. Betancourt-Martinez, P. Beiersdorfer, G.V. Brown, R.L. Kelley, C.A. Kilbourne, D. Koutroumpa, M.A. Leutenegger, F.S. Porter, Observation of highly disparate K-shell x-ray spectra produced by charge exchange with bare mid-Z ions. Phys. Rev. A **90**(5), 052723 (2014). https://doi.org/10.1103/PhysRevA.90.052723

A. Bhardwaj, R.F. Elsner, J.H. Waite Jr., G.R. Gladstone, T.E. Cravens, P.G. Ford, Chandra observation of an X-ray flare at Saturn: evidence of direct solar control on Saturn's disk X-ray emissions. Astrophys. J. Lett. **624**, 121–124 (2005a). https://doi.org/10.1086/430521

A. Bhardwaj, R.F. Elsner, J.H. Waite Jr., G.R. Gladstone, T.E. Cravens, P.G. Ford, The discovery of oxygen $K\alpha$ X-ray emission from the rings of Saturn. Astrophys. J. Lett. **627**, 73–76 (2005b). https://doi.org/10.1086/431933

A. Bhardwaj, R.F. Elsner, G. Randall Gladstone, T.E. Cravens, C.M. Lisse, K. Dennerl, G. Branduardi-Raymont, B.J. Wargelin, J. Hunter Waite, I. Robertson, N. Østgaard, P. Beiersdorfer, S.L. Snowden, V. Kharchenko, X-rays from solar system objects. Planet. Space Sci. **55**, 1135–1189 (2007). https://doi.org/10.1016/j.pss.2006.11.009

L. Billingham, S.J. Schwartz, D.G. Sibeck, The statistics of foreshock cavities: results of a Cluster survey. Ann. Geophys. **26**, 3653–3667 (2008). https://doi.org/10.5194/angeo-26-3653-2008

L. Billingham, S.J. Schwartz, M. Wilber, Foreshock cavities and internal foreshock boundaries. Planet. Space Sci. **59**, 456–467 (2011). https://doi.org/10.1016/j.pss.2010.01.012

R.D. Blandford, J.P. Ostriker, Particle acceleration by astrophysical shocks. Astrophys. J. Lett. **221**, 29–32 (1978). https://doi.org/10.1086/182658

D. Bodewits, D.J. Christian, M. Torney, M. Dryer, C.M. Lisse, K. Dennerl, T.H. Zurbuchen, S.J. Wolk, A.G.G.M. Tielens, R. Hoekstra, Spectral analysis of the Chandra comet survey. Astron. Astrophys. **469**, 1183–1195 (2007). https://doi.org/10.1051/0004-6361:20077410

B.R. Boller, H.L. Stolov, Explorer 18 study of the stability of the magnetopause using a Kelvin-Helmholtz instability criterion. J. Geophys. Res. **78**, 8078 (1973). https://doi.org/10.1029/JA078i034p08078

J.E. Borovsky, M.H. Denton, Solar wind turbulence and shear: a superposed-epoch analysis of corotating interaction regions at 1 AU. J. Geophys. Res. Space Phys. **115**, 10101 (2010). https://doi.org/10.1029/2009JA014966

A. Bößwetter, S. Simon, T. Bagdonat, U. Motschmann, M. Fränz, E. Roussos, N. Krupp, J. Woch, J. Schüle, S. Barabash, R. Lundin, Comparison of plasma data from ASPERA-3/Mars-Express with a 3-D hybrid simulation. Ann. Geophys. **25**, 1851–1864 (2007). https://doi.org/10.5194/angeo-25-1851-2007

C.S. Bowyer, G.B. Field, J.E. Mack, Detection of an anisotropic soft X-ray background flux. Nature **217**, 32–34 (1968). https://doi.org/10.1038/217032a0

L.H. Brace, R.F. Theis, W.R. Hoegy, Plasma clouds above the ionopause of Venus and their implications. Planet. Space Sci. **30**, 29–37 (1982). https://doi.org/10.1016/0032-0633(82)90069-1

L.H. Brace, R.F. Theis, J.D. Mihalov, Response of nightside ionosphere and ionotail of Venus to variations in solar EUV and solar wind dynamic pressure. J. Geophys. Res. **95**, 4075–4084 (1990). https://doi.org/10.1029/JA095iA04p04075

L.H. Brace, R.F. Theis, W.R. Hoegy, J.H. Wolfe, J.D. Mihalov, C.T. Russell, R.C. Elphic, A.F. Nagy, The dynamic behavior of the Venus ionosphere in response to solar wind interactions. J. Geophys. Res. **85**, 7663–7678 (1980). https://doi.org/10.1029/JA085iA13p07663

D.A. Brain, Mars Global Surveyor measurements of the Martian solar wind interaction. Space Sci. Rev. **126**, 77–112 (2006). https://doi.org/10.1007/s11214-006-9122-x

D.A. Brain, J.S. Halekas, R. Lillis, D.L. Mitchell, R.P. Lin, D.H. Crider, Variability of the altitude of the Martian sheath. Geophys. Res. Lett. **32**, 18203 (2005). https://doi.org/10.1029/2005GL023126

D.A. Brain, A.H. Baker, J. Briggs, J.P. Eastwood, J.S. Halekas, T.-D. Phan, Episodic detachment of Martian crustal magnetic fields leading to bulk atmospheric plasma escape. Geophys. Res. Lett. **37**, 14108 (2010a). https://doi.org/10.1029/2010GL043916

D. Brain, S. Barabash, A. Boesswetter, S. Bougher, S. Brecht, G. Chanteur, D. Hurley, E. Dubinin, X. Fang, M. Fraenz, J. Halekas, E. Harnett, M. Holmstrom, E. Kallio, H. Lammer, S. Ledvina, M. Liemohn, K. Liu, J. Luhmann, Y. Ma, R. Modolo, A. Nagy, U. Motschmann, H. Nilsson, H. Shinagawa, S. Simon, N. Terada, A comparison of global models for the solar wind interaction with Mars. Icarus **206**, 139–151 (2010b). https://doi.org/10.1016/j.icarus.2009.06.030

G. Branduardi-Raymont, C. Wang (Smile Team), SMILE (Solar wind Magnetosphere Ionosphere Link Explorer): X-ray imaging of the Sun-Earth connection, in *XMM-Newton: The Next Decade* (2016), p. 81

G. Branduardi-Raymont, R.F. Elsner, G.R. Gladstone, G. Ramsay, P. Rodriguez, R. Soria, J.H. Waite Jr., First observation of Jupiter by XMM-Newton. Astron. Astrophys. **424**, 331–337 (2004). https://doi.org/10.1051/0004-6361:20041149

G. Branduardi-Raymont, A. Bhardwaj, R.F. Elsner, G.R. Gladstone, G. Ramsay, P. Rodriguez, R. Soria, J.H. Waite Jr., T.E. Cravens, A study of Jupiter's aurorae with XMM-Newton. Astron. Astrophys. **463**, 761–774 (2007). https://doi.org/10.1051/0004-6361:20066406

G. Branduardi-Raymont, R.F. Elsner, M. Galand, D. Grodent, T.E. Cravens, P. Ford, G.R. Gladstone, J.H. Waite, Spectral morphology of the X-ray emission from Jupiter's aurorae. J. Geophys. Res. Space Phys. **113**, 02202 (2008). https://doi.org/10.1029/2007JA012600

G. Branduardi-Raymont, A. Bhardwaj, R.F. Elsner, P. Rodriguez, X-rays from Saturn: a study with XMM-Newton and Chandra over the years 2002–05. Astron. Astrophys. **510**, 73 (2010). https://doi.org/10.1051/0004-6361/200913110

G. Branduardi-Raymont, P.G. Ford, K.C. Hansen, L. Lamy, A. Masters, B. Cecconi, A.J. Coates, M.K. Dougherty, G.R. Gladstone, P. Zarka, Search for Saturn's X-ray aurorae at the arrival of a solar wind shock. J. Geophys. Res. Space Phys. **118**, 2145–2156 (2013). https://doi.org/10.1002/jgra.50112

S.H. Brecht, S.A. Ledvina, The solar wind interaction with the Martian ionosphere/atmosphere. Space Sci. Rev. **126**, 15–38 (2006). https://doi.org/10.1007/s11214-006-9084-z

E.J. Bunce, S.W.H. Cowley, T.K. Yeoman, Jovian cusp processes: implications for the polar aurora. J. Geophys. Res. Space Phys. **109**, 9–13 (2004). https://doi.org/10.1029/2003JA010280

J.L. Burch, Rate of erosion of dayside magnetic flux based on a quantitative study of the dependence of polar cusp latitude on the interplanetary magnetic field. Radio Sci. **8**, 955–961 (1973). https://doi.org/10.1029/RS008i011p00955

J.L. Burch, J. Goldstein, B.R. Sandel, Cause of plasmasphere corotation lag. Geophys. Res. Lett. **31**, 05802 (2004). https://doi.org/10.1029/2003GL019164

L.F. Burlaga, N.F. Ness, Tangential discontinuities in the solar wind. Sol. Phys. **9**, 467–477 (1969). https://doi.org/10.1007/BF02391672

 Springer

I.H. Cairns, C.L. Grabbe, Towards an MHD theory for the standoff distance of Earth's bow shock. Geophys. Res. Lett. **21**, 2781–2784 (1994). https://doi.org/10.1029/94GL02551

I.H. Cairns, J.G. Lyon, Magnetic field orientation effects on the standoff distance of Earth's bow shock. Geophys. Res. Lett. **23**, 2883–2886 (1996). https://doi.org/10.1029/96GL02755

M. Candidi, C.-I. Meng, Low-altitude observations of the conjugate polar cusps. J. Geophys. Res. **93**, 923–931 (1988). https://doi.org/10.1029/JA093iA02p00923

J.F. Carbary, C.I. Meng, Correlation of cusp latitude with B_z and AE (12) using nearly one year's data. J. Geophys. Res. **91**, 10047–10054 (1986). https://doi.org/10.1029/JA091iA09p10047

D.L. Carpenter, R.R. Anderson, An ISEE/Whistler model of equatorial electron density in the magnetosphere. J. Geophys. Res. **97**, 1097–1108 (1992). https://doi.org/10.1029/91JA01548

J.A. Carter, S. Sembay, Identifying XMM-Newton observations affected by solar wind charge exchange. Part I. Astron. Astrophys. **489**, 837–848 (2008). https://doi.org/10.1051/0004-6361:200809997

J.A. Carter, S. Sembay, A.M. Read, A high charge state coronal mass ejection seen through solar wind charge exchange emission as detected by XMM-Newton. Mon. Not. R. Astron. Soc. **402**, 867–878 (2010). https://doi.org/10.1111/j.1365-2966.2009.15985.x

J.A. Carter, S. Sembay, A.M. Read, Identifying XMM-Newton observations affected by solar wind charge exchange—Part II. Astron. Astrophys. **527**, 115 (2011). https://doi.org/10.1051/0004-6361/201015817

S.V. Chalov, Helium pickup ion focusing cone as an indicator of the interstellar flow direction. Mon. Not. R. Astron. Soc. **443**, 25–28 (2014). https://doi.org/10.1093/mnrasl/slu074

J.W. Chamberlain, Planetary coronae and atmospheric evaporation. Planet. Space Sci. **11**, 901–960 (1963). https://doi.org/10.1016/0032-0633(63)90122-3

E. Chané, J. Saur, F.M. Neubauer, J. Raeder, S. Poedts, Observational evidence of Alfvén wings at the Earth. J. Geophys. Res. Space Phys. **117**, 09217 (2012). https://doi.org/10.1029/2012JA017628

J.F. Chapman, I.H. Cairns, Modeling of Earth's bow shock: applications. J. Geophys. Res. Space Phys. **109**, 11202 (2004). https://doi.org/10.1029/2004JA010540

J.F. Chapman, I.H. Cairns, J.G. Lyon, C.R. Boshuizen, MHD simulations of Earth's bow shock: interplanetary magnetic field orientation effects on shape and position. J. Geophys. Res. Space Phys. **109**, 04215 (2004). https://doi.org/10.1029/2003JA010235

A. Chicarro, P. Martin, R. Trautner, The Mars Express mission: an overview, in *Mars Express: The Scientific Payload*, ed. by A. Wilson, A. Chicarro. ESA Special Publication, vol. 1240 (2004), pp. 3–13

S.P. Christon, D.C. Hamilton, G. Gloeckler, T.E. Eastmann, High charge state carbon and oxygen ions in Earth's equatorial quasi-trapping region. J. Geophys. Res. **99**, 13465 (1994). https://doi.org/10.1029/93JA03328

J.T. Clarke, J. Nichols, J.-C. Gérard, D. Grodent, K.C. Hansen, W. Kurth, G.R. Gladstone, J. Duval, S. Wannawichian, E. Bunce, S.W.H. Cowley, F. Crary, M. Dougherty, L. Lamy, D. Mitchell, W. Pryor, K. Retherford, T. Stallard, B. Zieger, P. Zarka, B. Cecconi, Response of Jupiter's and Saturn's auroral activity to the solar wind. J. Geophys. Res. Space Phys. **114**, 05210 (2009). https://doi.org/10.1029/2008JA013694

A.J. Coates, Heavy ion effects on cometary shocks. Adv. Space Res. **15**, 403–413 (1995). https://doi.org/10.1016/0273-1177(94)00125-K

I. Cohen, D.G. Mitchell, L.M. Kistler, B.H. Mauk, B.J. Anderson, J.H. Westlake, S. Ohtani, D.C. Hamilton, D.L. Turner, J.B. Blake, J.F. Fennell, A. Jaynes, T.W. Leonard, A. Gerrard, L.J. Lanzerotti, R.C. Allen, J.L. Burch, Dominance of high energy (> 150 keV) heavy ion intensities in Earth's middle to outer magnetosphere. J. Geophys. Res. **122** 9282–9293 (2017). https://doi.org/10.1002/2017JA024351

A. Colaprete, P. Schultz, J. Heldmann, D. Wooden, M. Shirley, K. Ennico, B. Hermalyn, W. Marshall, A. Ricco, R.C. Elphic, D. Goldstein, D. Summy, G.D. Bart, E. Asphaug, D. Korycansky, D. Landis, L. Sollitt, Detection of water in the LCROSS ejecta plume. Science **330**, 463 (2010). https://doi.org/10.1126/science.1186986

A. Colaprete, M. Sarantos, D.H. Wooden, T.J. Stubbs, A.M. Cook, M. Shirley, How surface composition and meteoroid impacts mediate sodium and potassium in the lunar exosphere. Science **351**, 249–252 (2016). https://doi.org/10.1126/science.aad2380

L. Colin, The Pioneer Venus program. J. Geophys. Res. **85**, 7575–7598 (1980). https://doi.org/10.1029/JA085iA13p07575

M.R. Collier, T.E. Moore, M.-C. Fok, B. Pilkerton, S. Boardsen, H. Khan, Low-energy neutral atom signatures of magnetopause motion in response to southward B_z. J. Geophys. Res. Space Phys. **110**, 02102 (2005a). https://doi.org/10.1029/2004JA010626

M.R. Collier, T.E. Moore, S.L. Snowden, K.D. Kuntz, One-up on L1: can X-rays provide longer advanced warning of solar wind flux enhancements than upstream monitors? Adv. Space Res. **35**, 2157–2161 (2005b). https://doi.org/10.1016/j.asr.2005.02.035

M.R. Collier, F.S. Porter, D.G. Sibeck, J.A. Carter, M.P. Chiao, D.J. Chornay, T. Cravens, M. Galeazzi, J.W. Keller, D. Koutroumpa, K. Kuntz, A.M. Read, I.P. Robertson, S. Sembay, S. Snowden, N. Thomas,

Prototyping a global soft X-ray imaging instrument for heliophysics, planetary science, and astrophysics science. Astron. Nachr. **333**, 378 (2012). https://doi.org/10.1002/asna.201211662

M.R. Collier, S.L. Snowden, M. Sarantos, M. Benna, J.A. Carter, T.E. Cravens, W.M. Farrell, S. Fatemi, H.K. Hills, R.R. Hodges, M. Holmström, K.D. Kuntz, F.S. Porter, A. Read, I.P. Robertson, S.F. Sembay, D.G. Sibeck, T.J. Stubbs, P. Travnicek, B.M. Walsh, On lunar exospheric column densities and solar wind access beyond the terminator from ROSAT soft X-ray observations of solar wind charge exchange. J. Geophys. Res., Planets **119**, 1459–1478 (2014). https://doi.org/10.1002/2014JE004628

M.R. Collier, F.S. Porter, D.G. Sibeck, J.A. Carter, M.P. Chiao, D.J. Chornay, T.E. Cravens, M. Galeazzi, J.W. Keller, D. Koutroumpa, J. Kujawski, K. Kuntz, A.M. Read, I.P. Robertson, S. Sembay, S.L. Snowden, N. Thomas, Y. Uprety, B.M. Walsh, Invited article: first flight in space of a wide-field-of-view soft x-ray imager using lobster-eye optics: instrument description and initial flight results. Rev. Sci. Instrum. **86**(7), 071301 (2015). https://doi.org/10.1063/1.4927259

G.A. Collinson, D.G. Sibeck, N. Shane, T.L. Zhang, A. Fedorov, S. Barabash, A.J. Coates, T.E. Moore, J.A. Slavin, V.M. Uritsky, S. Boardsen, M. Sarantos, A survey of hot flow anomalies at Venus. J. Geophys. Res. Space Phys. **119**, 978–991 (2014). https://doi.org/10.1002/2013JA018863

G.A. Collinson, J. Grebowsky, D.G. Sibeck, L.K. Jian, S. Boardsen, J. Espley, D. Hartle, T.L. Zhang, S. Barabash, Y. Futaana, P. Kollmann, The impact of a slow interplanetary coronal mass ejection on Venus. J. Geophys. Res. Space Phys. **120**, 3489–3502 (2015). https://doi.org/10.1002/2014JA020616

G.A. Collinson, R.A. Frahm, A. Glocer, A.J. Coates, J.M. Grebowsky, S. Barabash, S.D. Domagal-Goldman, A. Fedorov, Y. Futaana, L.K. Gilbert, G. Khazanov, T.A. Nordheim, D. Mitchell, T.E. Moore, W.K. Peterson, J.D. Winningham, T.L. Zhang, The electric wind of Venus: a global and persistent "polar wind"-like ambipolar electric field sufficient for the direct escape of heavy ionospheric ions. Geophys. Res. Lett. **43**, 5926–5934 (2016). https://doi.org/10.1002/2016GL068327

G. Collinson, D. Mitchell, A. Glocer, J. Grebowsky, W.K. Peterson, J. Connerney, L. Andersson, J. Espley, C. Mazelle, J.-A. Sauvaud, A. Fedorov, Y. Ma, S. Bougher, R. Lillis, R. Ergun, B. Jakosky, Electric Mars: the first direct measurement of an upper limit for the Martian "polar wind" electric potential. Geophys. Res. Lett. **42**, 9128–9134 (2015). https://doi.org/10.1002/2015GL065084

J.C. Cook, S.A. Stern, Sporadic increases in lunar atmospheric helium detected by LAMP. Icarus **236**, 48–55 (2014). https://doi.org/10.1016/j.icarus.2014.02.001

T.E. Cravens, Comet Hyakutake x-ray source: charge transfer of solar wind heavy ions. Geophys. Res. Lett. **24**, 105–108 (1997). https://doi.org/10.1029/96GL03780

T.E. Cravens, Heliospheric X-ray emission associated with charge transfer of the solar wind with interstellar neutrals. Astrophys. J. Lett. **532**, 153–156 (2000a). https://doi.org/10.1086/312574

T.E. Cravens, X-ray emission from comets and planets. Adv. Space Res. **26**, 1443–1451 (2000b). https://doi.org/10.1016/S0273-1177(00)00100-9

T.E. Cravens, X-ray emission from comets. Science **296**, 1042–1046 (2002). https://doi.org/10.1126/science.1070001

T.E. Cravens, A.N. Maurellis, X-ray emission from scattering and fluorescence of solar X-rays at Venus and Mars. Geophys. Res. Lett. **28**, 3043–3046 (2001). https://doi.org/10.1029/2001GL013021

T.E. Cravens, I.P. Robertson, S.L. Snowden, Temporal variations of geocoronal and heliospheric X-ray emission associated with the solar wind interaction with neutrals. J. Geophys. Res. **106**, 24883–24892 (2001). https://doi.org/10.1029/2000JA000461

T.E. Cravens, E. Howell, J.H. Waite, G.R. Gladstone, Auroral oxygen precipitation at Jupiter. J. Geophys. Res. **100**, 17153–17162 (1995). https://doi.org/10.1029/95JA00970

T.E. Cravens, J.H. Waite, T.I. Gombosi, N. Lugaz, G.R. Gladstone, B.H. Mauk, R.J. MacDowall, Implications of Jovian X-ray emission for magnetosphere-ionosphere coupling. J. Geophys. Res. Space Phys. **108**, 1465 (2003). https://doi.org/10.1029/2003JA010050

D.H. Crider, J. Espley, D.A. Brain, D.L. Mitchell, J.E.P. Connerney, M.H. AcuñA, Mars Global Surveyor observations of the Halloween 2003 solar superstorm's encounter with Mars. J. Geophys. Res. Space Phys. **110**, 9–21 (2005). https://doi.org/10.1029/2004JA010881

N.U. Crooker, Dayside merging and cusp geometry. J. Geophys. Res. **84**, 951–959 (1979). https://doi.org/10.1029/JA084iA03p00951

F. Dalaudier, J.L. Bertaux, V.G. Kurt, E.N. Mironova, Characteristics of interstellar helium observed with Prognoz 6 58.4-nm photometers. Astron. Astrophys. **134**, 171–184 (1984)

F. Darrouzet, D.L. Gallagher, N. André, D.L. Carpenter, I. Dandouras, P.M.E. Décréau, J. de Keyser, R.E. Denton, J.C. Foster, J. Goldstein, M.B. Moldwin, B.W. Reinisch, B.R. Sandel, J. Tu, Plasmaspheric density structures and dynamics: properties observed by the CLUSTER and IMAGE missions. Space Sci. Rev. **145**, 55–106 (2009). https://doi.org/10.1007/s11214-008-9438-9

R.B. Decker, Formation of shock-spike events at quasi-perpendicular shocks. J. Geophys. Res. **88**, 9959–9973 (1983). https://doi.org/10.1029/JA088iA12p09959

K. Dennerl, Discovery of X-rays from Mars with Chandra. Astron. Astrophys. **394**, 1119–1128 (2002). https://doi.org/10.1051/0004-6361:20021116

K. Dennerl, X-rays from Mars. Space Sci. Rev. **126**, 403–433 (2006). https://doi.org/10.1007/s11214-006-9028-7

K. Dennerl, X-rays from Venus observed with Chandra. Planet. Space Sci. **56**, 1414–1423 (2008). https://doi.org/10.1016/j.pss.2008.03.008

K. Dennerl, Charge transfer reactions. Space Sci. Rev. **157**, 57–91 (2010). https://doi.org/10.1007/s11214-010-9720-5

K. Dennerl, J. Englhauser, J. Trümper, X-ray emissions from comets detected in the Röntgen X-ray satellite all-sky survey. Science **277**, 1625–1630 (1997). https://doi.org/10.1126/science.277.5332.1625

K. Dennerl, V. Burwitz, J. Englhauser, C. Lisse, S. Wolk, Discovery of X-rays from Venus with Chandra. Astron. Astrophys. **386**, 319–330 (2002). https://doi.org/10.1051/0004-6361:20020097

K. Dennerl, C.M. Lisse, A. Bhardwaj, V. Burwitz, J. Englhauser, H. Gunell, M. Holmström, F. Jansen, V. Kharchenko, P.M. Rodríguez-Pascual, First observation of Mars with XMM-Newton. High resolution X-ray spectroscopy with RGS. Astron. Astrophys. **451**, 709–722 (2006). https://doi.org/10.1051/0004-6361:20054253

K. Dennerl, C.M. Lisse, A. Bhardwaj, D.J. Christian, S.J. Wolk, D. Bodewits, T.H. Zurbuchen, M. Combi, S. Lepri, Solar system X-rays from charge exchange processes. Astron. Nachr. **333**, 324 (2012). https://doi.org/10.1002/asna.201211663

A.P. Dimmock, K. Nykyri, The statistical mapping of magnetosheath plasma properties based on THEMIS measurements in the magnetosheath interplanetary medium reference frame. J. Geophys. Res. Space Phys. **118**, 4963–4976 (2013). https://doi.org/10.1002/jgra.50465

A.P. Dimmock, T.I. Pulkkinen, A. Osmane, K. Nykyri, The dawn-dusk asymmetry of ion density in the dayside magnetosheath and its annual variability measured by THEMIS. Ann. Geophys. **34**, 511–528 (2016)

A. Dmitriev, A. Suvorova, Equatorial trench at the magnetopause under saturation. J. Geophys. Res. Space Phys. **117**, 8226 (2012). https://doi.org/10.1029/2012JA017834

A. Dmitriev, A. Suvorova, J.-K. Chao, A predictive model of geosynchronous magnetopause crossings. J. Geophys. Res. Space Phys. **116**, 5208 (2011). https://doi.org/10.1029/2010JA016208

A.V. Dmitriev, A.V. Suvorova, Three-dimensional artificial neural network model of the dayside magnetopause. J. Geophys. Res. **105**, 18909 (2000). https://doi.org/10.1029/2000JA900008

A.V. Dmitriev, A.V. Suvorova, J.K. Chao, Y.-H. Yang, Dawn-dusk asymmetry of geosynchronous magnetopause crossings. J. Geophys. Res. Space Phys. **109**, 5203 (2004). https://doi.org/10.1029/2003JA010171

A. Dmitriev, J.-K. Chao, M. Thomsen, A. Suvorova, Geosynchronous magnetopause crossings on 29–31 October 2003. J. Geophys. Res. Space Phys. **110**, 8209 (2005). https://doi.org/10.1029/2004JA010582

J.F. Drake, M. Swisdak, K.M. Schoeffler, B.N. Rogers, S. Kobayashi, Formation of secondary islands during magnetic reconnection. Geophys. Res. Lett. **33**, 13105 (2006). https://doi.org/10.1029/2006GL025957

J. Dreher, G.T. Birk, K. Schindler, A. Otto, Role of magnetic reconnection in Venus ionopause activity. J. Geophys. Res. **100**, 14833–14842 (1995). https://doi.org/10.1029/95JA00522

E.M. Dubinin, K. Sauer, R. Lundin, K. Baumgärtel, A. Bogdanov, Structuring of the transition region (plasma mantle) of the Martian magnetosphere. Geophys. Res. Lett. **23**, 785–788 (1996). https://doi.org/10.1029/96GL00701

E. Dubinin, K. Sauer, M. Delva, T. Tanaka, The IMF control of the Martian bow shock and plasma flow in the magnetosheath. Predictions of 3-D simulations and observations. Earth Planets Space **50**, 873–882 (1998)

W.R. Dunn, G. Branduardi-Raymont, R.F. Elsner, M.F. Vogt, L. Lamy, P.G. Ford, A.J. Coates, G.R. Gladstone, C.M. Jackman, J.D. Nichols, I.J. Rae, A. Varsani, T. Kimura, K.C. Hansen, J.M. Jasinski, The impact of an ICME on the Jovian X-ray aurora. J. Geophys. Res. Space Phys. **121**, 2274–2307 (2016). https://doi.org/10.1002/2015JA021888

F. Duru, D.A. Gurnett, R.A. Frahm, J.D. Winningham, D.D. Morgan, G.G. Howes, Steep, transient density gradients in the Martian ionopshere similar to the ionopause at Venus. AGU Fall Meeting Abstracts (2009)

Š. Dušík, G. Granko, J. Šafránková, Z. Němeček, K. Jelínek, IMF cone angle control of the magnetopause location: statistical study. Geophys. Res. Lett. **37**, 19103 (2010). https://doi.org/10.1029/2010GL044965

R.W. Ebert, D.J. McComas, H.A. Elliott, R.J. Forsyth, J.T. Gosling, Bulk properties of the slow and fast solar wind and interplanetary coronal mass ejections measured by Ulysses: three polar orbits of observations. J. Geophys. Res. Space Phys. **114**, 1109 (2009). https://doi.org/10.1029/2008JA013631

N.J.T. Edberg, M. Lester, S.W.H. Cowley, D.A. Brain, M. Fränz, S. Barabash, Magnetosonic Mach number effect of the position of the bow shock at Mars in comparison to Venus. J. Geophys. Res. Space Phys. **115**, 07203 (2010). https://doi.org/10.1029/2009JA014998

N.J.T. Edberg, H. Nilsson, Y. Futaana, G. Stenberg, M. Lester, S.W.H. Cowley, J.G. Luhmann, T.R. McEnulty, H.J. Opgenoorth, A. Fedorov, S. Barabash, T.L. Zhang, Atmospheric erosion of Venus during stormy space weather. J. Geophys. Res. Space Phys. **116**, 09308 (2011). https://doi.org/10.1029/2011JA016749

R.C. Elphic, A.I. Ershkovich, On the stability of the ionopause of Venus. J. Geophys. Res. **89**, 997–1002 (1984). https://doi.org/10.1029/JA089iA02p00997

R.C. Elphic, G.T. Delory, B.P. Hine, P.R. Mahaffy, M. Horanyi, A. Colaprete, M. Benna, S.K. Noble, The lunar atmosphere and dust environment explorer mission. Space Sci. Rev. **185**, 3–25 (2014). https://doi.org/10.1007/s11214-014-0113-z

R.K. Elsen, R.M. Winglee, The average shape of the magnetopause: a comparison of three-dimensional global MHD and empirical models. J. Geophys. Res. **102**, 4799–4820 (1997). https://doi.org/10.1029/96JA03518

R.F. Elsner, N. Lugaz, J.H. Waite, T.E. Cravens, G.R. Gladstone, P. Ford, D. Grodent, A. Bhardwaj, R.J. MacDowall, M.D. Desch, T. Majeed, Simultaneous Chandra X ray, Hubble Space Telescope ultraviolet, and Ulysses radio observations of Jupiter's aurora. J. Geophys. Res. Space Phys. **110**, 01207 (2005). https://doi.org/10.1029/2004JA010717

N.V. Erkaev, C.J. Farrugia, H.K. Biernat, The role of the magnetic barrier in the solar wind-magnetosphere interaction. Planet. Space Sci. **51**, 745–755 (2003). https://doi.org/10.1016/S0032-0633(03)00111-9

A.I. Ershkovich, D.A. Mendis, On the penetration of the solar wind into the cometary ionosphere. Astrophys. J. **269**, 743–750 (1983). https://doi.org/10.1086/161083

C. Escoubet, J.M. Bosqued, The influence of IMF-Bz and/or AE on the polar cusp—an overview of observations from the Aureol-3 satellite. Planet. Space Sci. **37**, 609–626 (1989). https://doi.org/10.1016/0032-0633(89)90100-1

C.P. Escoubet, M.F. Smith, S.F. Fung, P.C. Anderson, R.A. Hoffman, E.M. Basinska, J.M. Bosqued, Staircase ion signature in the polar cusp—a case study. Geophys. Res. Lett. **19**, 1735–1738 (1992). https://doi.org/10.1029/92GL01806

Y. Ezoe, K. Ebisawa, N.Y. Yamasaki, K. Mitsuda, H. Yoshitake, N. Terada, Y. Miyoshi, R. Fujimoto, Time variability of the geocoronal solar-wind charge exchange in the direction of the celestial equator. Publ. Astron. Soc. Jpn. **62**, 981 (2010)

Y. Ezoe, Y. Miyoshi, H. Yoshitake, K. Mitsuda, N. Terada, S. Oishi, T. Ohashi, Enhancement of terrestrial diffuse X-ray emission associated with coronal mass ejection and geomagnetic storm. Publ. Astron. Soc. Jpn. **63**, 691–704 (2011). https://doi.org/10.1093/pasj/63.sp3.S691

Y. Ezoe, T. Kimura, S. Kasahara, A. Yamazaki, K. Mitsuda, M. Fujimoto, Y. Miyoshi, G. Branduardi-Raymont, K. Ishikawa, I. Mitsuishi, T. Ogawa, T. Kakiuchi, T. Ohashi, JUXTA: a new probe of X-ray emission from the Jupiter system. Adv. Space Res. **51**, 1605–1621 (2013). https://doi.org/10.1016/j.asr.2012.11.029

Y. Ezoe, I. Mitsuishi, U. Takagi, M. Koshiishi, K. Mitsuda, N.Y. Yamasaki, T. Ohashi, F. Kato, S. Sugiyama, R.E. Riveros, H. Yamaguchi, S. Fujihira, Y. Kanamori, K. Morishita, K. Nakajima, R. Maeda, Ultra light-weight and high-resolution X-ray mirrors using DRIE and X-ray LIGA techniques for space X-ray telescopes. Microsyst. Technol. **16**(8), 1633–1641 (2009). https://doi.org/10.1007/s00542-009-0981-4

G. Facskó, K. Kecskeméty, G. Erdős, M. Tátrallyay, P.W. Daly, I. Dandouras, A statistical study of hot flow anomalies using Cluster data. Adv. Space Res. **41**, 1286–1291 (2008). https://doi.org/10.1016/j.asr.2008.02.005

D.H. Fairfield, W. Baumjohann, G. Paschmann, H. Luehr, D.G. Sibeck, Upstream pressure variations associated with the bow shock and their effects on the magnetosphere. J. Geophys. Res. **95**, 3773–3786 (1990). https://doi.org/10.1029/JA095iA04p03773

D.H. Fairfield, I.H. Cairns, M.D. Desch, A. Szabo, A.J. Lazarus, M.R. Aellig, The location of low Mach number bow shocks at Earth. J. Geophys. Res. **106**, 25361–25376 (2001). https://doi.org/10.1029/2000JA000252

W.M. Farrell, T.J. Stubbs, J.S. Halekas, G.T. Delory, M.R. Collier, R.R. Vondrak, R.P. Lin, Loss of solar wind plasma neutrality and affect on surface potentials near the lunar terminator and shadowed polar regions. Geophys. Res. Lett. **35**, 05105 (2008). https://doi.org/10.1029/2007GL032653

M.H. Farris, C.T. Russell, Determining the standoff distance of the bow shock: Mach number dependence and use of models. J. Geophys. Res. **99**, 17 (1994). https://doi.org/10.1029/94JA01020

M.H. Farris, S.M. Petrinec, C.T. Russell, The thickness of the magnetosheath—constraints on the polytropic index. Geophys. Res. Lett. **18**, 1821–1824 (1991). https://doi.org/10.1029/91GL02090

C.J. Farrugia, N.V. Erkaev, H.K. Biernat, On the effects of solar wind dynamic pressure on the anisotropic terrestrial magnetosheath. J. Geophys. Res. **105**, 115–128 (2000). https://doi.org/10.1029/1999JA900350

C.J. Farrugia, N.V. Erkaev, H.K. Biernat, L.F. Burlaga, Anomalous magnetosheath properties during Earth passage of an interplanetary magnetic cloud. J. Geophys. Res. **100**, 19245–19258 (1995). https://doi.org/10.1029/95JA01080

S. Fatemi, M. Holmström, Y. Futaana, The effects of lunar surface plasma absorption and solar wind temperature anisotropies on the solar wind proton velocity space distributions in the low-altitude lunar plasma wake. J. Geophys. Res. Space Phys. **117**, 10105 (2012). https://doi.org/10.1029/2011JA017353

P.D. Feldman, D.M. Hurley, K.D. Retherford, G.R. Gladstone, S.A. Stern, W. Pryor, J.W. Parker, D.E. Kaufmann, M.W. Davis, M.H. Versteeg (LAMP Team), Temporal variability of lunar exospheric helium during January 2012 from LRO/LAMP. Icarus **221**, 854–858 (2012). https://doi.org/10.1016/j.icarus.2012.09.015

K.R. Flammer, The global interaction of comets with the solar wind, in *IAU Colloq. 116: Comets in the Post-Halley Era*, ed. by R.L. Newburn Jr., M. Neugebauer, J. Rahe. Astrophysics and Space Science Library, vol. 167 (1991) pp. 1125–1144. https://doi.org/10.1007/978-0-7923-1165-247

J.M. Forbes, F.G. Lemoine, S.L. Bruinsma, M.D. Smith, X. Zhang, Solar flux variability of Mars' exosphere densities and temperatures. Geophys. Res. Lett. **35**, 01201 (2008). https://doi.org/10.1029/2007GL031904

L.A. Frank, J.D. Craven, K.L. Ackerson, M.R. English, R.H. Eather, R.L. Carovillano, Global auroral imaging instrumentation for the Dynamics Explorer mission. Space Sci. Instrum. **5**, 369–393 (1981)

L.A. Frank, J.B. Sigwarth, J.D. Craven, J.P. Cravens, J.S. Dolan, M.R. Dvorsky, P.K. Hardebeck, J.D. Harvey, D.W. Muller, The Visible Imaging System (VIS) for the Polar spacecraft. Space Sci. Rev. **71**, 297–328 (1995). https://doi.org/10.1007/BF00751334

G.W. Fraser, J.E. Lees, J.F. Pearson, M.R. Sims, K. Roxburgh, X-ray focusing using microchannel plates, in *Multilayer and Grazing Incidence X-Ray/EUV Optics*, ed. by R.B. Hoover. Society of Photo-Optical Instrumentation Engineers (SPIE) Conference Series, vol. 1546 (1992), pp. 41–52

G.W. Fraser, J.D. Carpenter, D.A. Rothery, J.F. Pearson, A. Martindale, J. Huovelin, J. Treis, M. Anand, M. Anttila, M. Ashcroft, J. Benkoff, P. Bland, A. Bowyer, A. Bradley, J. Bridges, C. Brown, C. Bulloch, E.J. Bunce, U. Christensen, M. Evans, R. Fairbend, M. Feasey, F. Giannini, S. Hermann, M. Hesse, M. Hilchenbach, T. Jorden, K. Joy, M. Kaipiainen, I. Kitchingman, P. Lechner, G. Lutz, A. Malkki, K. Muinonen, J. Näränen, P. Portin, M. Prydderch, J.S. Juan, E. Sclater, E. Schyns, T.J. Stevenson, L. Strüder, M. Syrjasuo, D. Talboys, P. Thomas, C. Whitford, S. Whitehead, The Mercury imaging X-ray spectrometer (MIXS) on Bepi-Colombo. Planet. Space Sci. **58**, 79–95 (2010). https://doi.org/10.1016/j.pss.2009.05.004

H.U. Frey, S.B. Mende, S.A. Fuselier, T.J. Immel, N. ØStgaard, Proton aurora in the cusp during southward IMF. J. Geophys. Res. Space Phys. **108**, 1277 (2003). https://doi.org/10.1029/2003JA009861

M.J. Freyberg, On the zero-level of the soft X-ray background, in *IAU Colloq. 166: The Local Bubble and Beyond*, ed. by D. Breitschwerdt, M.J. Freyberg, J. Truemper. Lecture Notes in Physics, vol. 506 (Springer, Berlin, 1998), pp. 113–116. https://doi.org/10.1007/BFb0104704

R. Fujimoto, K. Mitsuda, D. Mccammon, Y. Takei, M. Bauer, Y. Ishisaki, S.F. Porter, H. Yamaguchi, K. Hayashida, N.Y. Yamasaki, Evidence for solar-wind charge-exchange X-ray emission from the Earth's magnetosheath. Publ. Astron. Soc. Jpn. **59**, 133–140 (2007). https://doi.org/10.1093/pasj/59.sp1.S133

S. Fung, High-altitude cusp positions sampled by the Hawkeye satellite. Phys. Chem. Earth **22**, 653–662 (1997). https://doi.org/10.1016/S0079-1946(97)88121-9

H.O. Funsten, F. Allegrini, P. Bochsler, G. Dunn, S. Ellis, D. Everett, M.J. Fagan, S.A. Fuselier, M. Granoff, M. Gruntman, A.A. Guthrie, J. Hanley, R.W. Harper, D. Heirtzler, P. Janzen, K.H. Kihara, B. King, H. Kucharek, M.P. Manzo, M. Maple, K. Mashburn, D.J. McComas, E. Moebius, J. Nolin, D. Piazza, S. Pope, D.B. Reisenfeld, B. Rodriguez, E.C. Roelof, L. Saul, S. Turco, P. Valek, S. Weidner, P. Wurz, S. Zaffke, The Interstellar Boundary Explorer High Energy (IBEX-Hi) neutral atom imager. Space Sci. Rev. **146**, 75–103 (2009). https://doi.org/10.1007/s11214-009-9504-y

S.A. Fuselier, S.M. Petrinec, K.J. Trattner, Stability of the high-latitude reconnection site for steady northward IMF. Geophys. Res. Lett. **27**, 473–476 (2000). https://doi.org/10.1029/1999GL003706

S.A. Fuselier, H.U. Frey, K.J. Trattner, S.B. Mende, J.L. Burch, Cusp aurora dependence on interplanetary magnetic field B_z. J. Geophys. Res. Space Phys. **107**, 1111 (2002). https://doi.org/10.1029/2001JA900165

S.A. Fuselier, H.O. Funsten, D. Heirtzler, P. Janzen, H. Kucharek, D.J. McComas, E. Möbius, T.E. Moore, S.M. Petrinec, D.B. Reisenfeld, N.A. Schwadron, K.J. Trattner, P. Wurz, Energetic neutral atoms from the Earth's subsolar magnetopause. Geophys. Res. Lett. **37**, 13101 (2010). https://doi.org/10.1029/2010GL044140

Y. Futaana, J.-Y. Chaufray, H.T. Smith, P. Garnier, H. Lichtenegger, M. Delva, H. Gröller, A. Mura, Exospheres and energetic neutral atoms of Mars, Venus and Titan. Space Sci. Rev. **162**, 213–266 (2011). https://doi.org/10.1007/s11214-011-9834-4

M. Galeazzi, M. Chiao, M.R. Collier, T. Cravens, D. Koutroumpa, K.D. Kuntz, S. Lepri, D. McCammon, F.S. Porter, K. Prasai, I. Robertson, S. Snowden, Y. Uprety, DXL: a sounding rocket mission for the study of solar wind charge exchange and local hot bubble X-ray emission. Exp. Astron. **32**, 83–99 (2011). https://doi.org/10.1007/s10686-011-9249-y

M. Galeazzi, M.R. Collier, T. Cravens, D. Koutroumpa, K.D. Kuntz, S. Lepri, D. McCammon, F.S. Porter, K. Prasai, I. Robertson, S. Snowden, N.E. Thomas, Y. Uprety, Solar wind charge exchange and local hot bubble X-ray emission with the DXL sounding rocket experiment. Astron. Nachr. **333**, 383 (2012). https://doi.org/10.1002/asna.201211665

M. Galeazzi, M. Chiao, M.R. Collier, T. Cravens, D. Koutroumpa, K.D. Kuntz, R. Lallement, S.T. Lepri, D. McCammon, K. Morgan, F.S. Porter, I.P. Robertson, S.L. Snowden, N.E. Thomas, Y. Uprety, E. Ursino, B.M. Walsh, The origin of the local 1/4-keV X-ray flux in both charge exchange and a hot bubble. Nature **512**, 171–173 (2014). https://doi.org/10.1038/nature13525

A.A. Galeev, K.I. Gringauz, S.I. Klimov, A.P. Remizov, R.Z. Sagdeev, S.P. Savin, A.Y. Sokolov, M.I. Verigin, K. Szegö, M. Tátrallyay, R. Grard, Y.G. Yeroshenko, M. Mogilevsky, W. Riedler, K. Schwingenschuh, Physical processes in the vicinity of the cometopause interpreted on the basis of plasma, magnetic field, and plasma wave data measured on board the Vega 2 spacecraft. J. Geophys. Res. **93**, 7527–7531 (1988). https://doi.org/10.1029/JA093iA07p07527

A.B. Galvin, Minor ion composition in CME-related solar wind, in coronal mass ejections, in *Geophys. Monogr. Series*, vol. 99, ed. by N. Crooker, J.A. Joselyn, J. Feynman (1997), pp. 253–260

H. Gao, V.H.S. Kwong, Charge transfer of O^{5+} and O^{4+} with CO at keV energies. Astrophys. J. **567**, 1272–1275 (2002). https://doi.org/10.1086/338589

J. Geiss, M. Witte, Properties of the interstellar gas inside the heliosphere. Space Sci. Rev. **78**, 229–238 (1996). https://doi.org/10.1007/BF00170809

H.B. Gilbody, Measurements of charge transfer and ionization in collisions involving hydrogen atoms. Adv. At. Mol. Phys. **22**, 143–195 (1986). https://doi.org/10.1016/S0065-2199(08)60336-X

G.R. Gladstone, J.H. Waite, D. Grodent, W.S. Lewis, F.J. Crary, R.F. Elsner, M.C. Weisskopf, T. Majeed, J.-M. Jahn, A. Bhardwaj, J.T. Clarke, D.T. Young, M.K. Dougherty, S.A. Espinosa, T.E. Cravens, A pulsating auroral X-ray hot spot on Jupiter. Nature **415**, 1000–1003 (2002)

G.R. Gladstone, S.A. Stern, K.D. Retherford, R.K. Black, D.C. Slater, M.W. Davis, M.H. Versteeg, K.B. Persson, J.W. Parker, D.E. Kaufmann, A.F. Egan, T.K. Greathouse, P.D. Feldman, D. Hurley, W.R. Pryor, A.R. Hendrix, LAMP: the Lyman alpha mapping project on NASA's lunar reconnaissance orbiter mission. Space Sci. Rev. **150**, 161–181 (2010). https://doi.org/10.1007/s11214-009-9578-6

G. Gloeckler, F.M. Ipavich, D.C. Hamilton, B. Wilken, W. Stuedemann, Solar wind carbon, nitrogen and oxygen abundances measured in the Earth's magnetosheath with AMPTE/CCE. Geophys. Res. Lett. **13**, 793–796 (1986). https://doi.org/10.1029/GL013i008p00793

G. Gloeckler, J. Cain, F.M. Ipavich, E.O. Tums, P. Bedini, L.A. Fisk, T.H. Zurbuchen, P. Bochsler, J. Fischer, R.F. Wimmer-Schweingruber, J. Geiss, R. Kallenbach, Investigation of the composition of solar and interstellar matter using solar wind and pickup ion measurements with SWICS and SWIMS on the ACE spacecraft. Space Sci. Rev. **86**, 497–539 (1998). https://doi.org/10.1023/A:1005036131689

G. Gloeckler, E. Möbius, J. Geiss, M. Bzowski, S. Chalov, H. Fahr, D.R. McMullin, H. Noda, M. Oka, D. Ruciński, R. Skoug, T. Terasawa, R. von Steiger, A. Yamazaki, T. Zurbuchen, Observations of the helium focusing cone with pickup ions. Astron. Astrophys. **426**, 845–854 (2004). https://doi.org/10.1051/0004-6361:20035768

C. Goetz, C. Koenders, I. Richter, K. Altwegg, J. Burch, C. Carr, E. Cupido, A. Eriksson, C. Güttler, P. Henri, P. Mokashi, Z. Nemeth, H. Nilsson, M. Rubin, H. Sierks, B. Tsurutani, C. Vallat, M. Volwerk, K.-H. Glassmeier, First detection of a diamagnetic cavity at comet 67P/Churyumov-Gerasimenko. Astron. Astrophys. **588**, 24 (2016a). https://doi.org/10.1051/0004-6361/201527728

C. Goetz, C. Koenders, K.C. Hansen, J. Burch, C. Carr, A. Eriksson, D. Frühauff, C. Güttler, P. Henri, H. Nilsson, I. Richter, M. Rubin, H. Sierks, B. Tsurutani, M. Volwerk, K.H. Glassmeier, Structure and evolution of the diamagnetic cavity at comet 67P/Churyumov-Gerasimenko. Mon. Not. R. Astron. Soc. **462**, 459–467 (2016b). https://doi.org/10.1093/mnras/stw3148

T.I. Gombosi, Charge exchange avalanche at the cometopause. Geophys. Res. Lett. **14**, 1174–1177 (1987). https://doi.org/10.1029/GL014i011p01174

W.D. Gonzalez, F.S. Mozer, A quantitative model for the potential resulting from reconnection with an arbitrary interplanetary magnetic field. J. Geophys. Res. **79**, 4186–4194 (1974). https://doi.org/10.1029/JA079i028p04186

P. Gorenstein, Focusing X-ray optics for astronomy. X-Ray Opt. Instrum. **2010**, 109740 (2010). Special issue on X-ray focusing: techniques and applications, id. 109740. https://doi.org/10.1155/2010/109740

J.T. Gosling, M.F. Thomsen, S.J. Bame, T.G. Onsager, C.T. Russell, The electron edge of the low latitude boundary layer during accelerated flow events. Geophys. Res. Lett. **17**, 1833–1836 (1990). https://doi.org/10.1029/GL017i011p01833

C.L. Grabbe, I.H. Cairns, Analytic MHD theory for Earth's bow shock at low Mach numbers. J. Geophys. Res. **100**, 19941–19950 (1995). https://doi.org/10.1029/95JA01286

C. Grava, J.-Y. Chaufray, K.D. Retherford, G.R. Gladstone, T.K. Greathouse, D.M. Hurley, R.R. Hodges, A.J. Bayless, J.C. Cook, S.A. Stern, Lunar exospheric argon modeling. Icarus **255**, 135–147 (2015). https://doi.org/10.1016/j.icarus.2014.09.029

J.B. Greenwood, I.D. Williams, S.J. Smith, A. Chutjian, Experimental investigation of the processes determining x-ray emission intensities from charge-exchange collisions. Phys. Rev. A **63**(6), 062707 (2001). https://doi.org/10.1103/PhysRevA.63.062707

K.I. Gringauz, T.I. Gombosi, A.P. Remizov, I. Apathy, I. Szemerey, M.I. Verigin, L.I. Denchikova, A.V. Dyachkov, E. Keppler, I.N. Klimenko, A.K. Richter, A.J. Somogyi, K. Szego, S. Szendro, M. Tatrallyay, A. Varga, G.A. Vladimirova, First in situ plasma and neutral gas measurements at comet Halley. Nature **321**, 282–285 (1986). https://doi.org/10.1038/321282a0

M.A. Gruntman, Neutral solar wind properties: advance warning of major geomagnetic storms. J. Geophys. Res. **99**, 19 (1994). https://doi.org/10.1029/94JA01571

H. Gunell, M. Holmström, E. Kallio, P. Janhunen, K. Dennerl, X rays from solar wind charge exchange at Mars: a comparison of simulations and observations. Geophys. Res. Lett. **31**, 22801 (2004). https://doi.org/10.1029/2004GL020953

H. Gunell, M. Holmström, E. Kallio, P. Janhunen, K. Dennerl, Simulations of X-rays from solar wind charge exchange at Mars: parameter dependence. Adv. Space Res. **36**, 2057–2065 (2005). https://doi.org/10.1016/j.asr.2005.06.007

H. Gunell, E. Kallio, R. Jarvinen, P. Janhunen, M. Holmström, K. Dennerl, Simulations of solar wind charge exchange X-ray emissions at Venus. Geophys. Res. Lett. **34**, 03107 (2007). https://doi.org/10.1029/2006GL028602

H. Gunell, U.V. Amerstorfer, H. Nilsson, C. Grima, M. Koepke, M. Fränz, J.D. Winningham, R.A. Frahm, J.-A. Sauvaud, A. Fedorov, N.V. Erkaev, H.K. Biernat, M. Holmström, R. Lundin, S. Barabash, Shear driven waves in the induced magnetosphere of Mars. Plasma Phys. Control. Fusion **50**(7), 074018 (2008). https://doi.org/10.1088/0741-3335/50/7/074018

J. Guo, X. Feng, J. Zhang, P. Zuo, C. Xiang, Statistical properties and geoefficiency of interplanetary coronal mass ejections and their sheaths during intense geomagnetic storms. J. Geophys. Res. Space Phys. **115**, 09107 (2010). https://doi.org/10.1029/2009JA015140

F. Haberl, The XMM-Newton survey of the large (and small) Magellanic cloud, in *The X-Ray Universe 2014* (2014), p. 4

J.S. Halekas, S.D. Bale, D.L. Mitchell, R.P. Lin, Electrons and magnetic fields in the lunar plasma wake. J. Geophys. Res. Space Phys. **110**, 07222 (2005). https://doi.org/10.1029/2004JA010991

E.M. Harnett, High-resolution multifluid simulations of flux ropes in the Martian magnetosphere. J. Geophys. Res. Space Phys. **114**, 01208 (2009). https://doi.org/10.1029/2008JA013648

E.M. Harnett, R.M. Winglee, The influence of a mini-magnetopause on the magnetic pileup boundary at Mars. Geophys. Res. Lett. **30**, 2074 (2003). https://doi.org/10.1029/2003GL017852

E.M. Harnett, R.M. Winglee, Three-dimensional fluid simulations of plasma asymmetries in the Martian magnetotail caused by the magnetic anomalies. J. Geophys. Res. Space Phys. **110**, 07226 (2005). https://doi.org/10.1029/2003JA010315

H. Hasegawa, M. Fujimoto, K. Maezawa, Y. Saito, T. Mukai, Geotail observations of the dayside outer boundary region: interplanetary magnetic field control and dawn-dusk asymmetry. J. Geophys. Res. Space Phys. **108**, 1163 (2003). https://doi.org/10.1029/2002JA009667

H. Hasegawa, A. Retinò, A. Vaivads, Y. Khotyaintsev, M. André, T.K.M. Nakamura, W.-L. Teh, B.U.Ö. Sonnerup, S.J. Schwartz, Y. Seki, M. Fujimoto, Y. Saito, H. Rème, P. Canu, Kelvin-Helmholtz waves at the Earth's magnetopause: multiscale development and associated reconnection. J. Geophys. Res. Space Phys. **114**, 12207 (2009). https://doi.org/10.1029/2009JA014042

W.J. Heikkila, J.D. Winningham, Penetration of magnetosheath plasma to low altitudes through the dayside magnetospheric cusps. J. Geophys. Res. **76**, 883 (1971). https://doi.org/10.1029/JA076i004p00883

D.B. Henley, R.L. Shelton, An XMM-Newton survey of the soft X-ray background. I. The O VII and O VIII lines between $l = 120°$ and $l = 240°$. Astrophys. J. Suppl. Ser. **187**, 388–408 (2010). https://doi.org/10.1088/0067-0049/187/2/388

D.B. Henley, R.L. Shelton, An XMM-Newton survey of the soft X-ray background. II. An all-sky catalog of diffuse O VII and O VIII emission intensities. Astrophys. J. Suppl. Ser. **202**, 14 (2012). https://doi.org/10.1088/0067-0049/202/2/14

M. Hesse, J. Birn, Three-dimensional MHD modeling of magnetotail dynamics for different polytropic indices. J. Geophys. Res. **97**, 3965–3976 (1992). https://doi.org/10.1029/91JA03003

H. Hietala, N. Partamies, T.V. Laitinen, L.B.N. Clausen, G. Facskó, A. Vaivads, H.E.J. Koskinen, I. Dandouras, H. Rème, E.A. Lucek, Supermagnetosonic subsolar magnetosheath jets and their effects: from the solar wind to the ionospheric convection. Ann. Geophys. **30**, 33–48 (2012). https://doi.org/10.5194/angeo-30-33-2012

R.R. Hodges, P.R. Mahaffy, Synodic and semiannual oscillations of argon-40 in the lunar exosphere. Geophys. Res. Lett. **43**, 22–27 (2016). https://doi.org/10.1002/2015GL067293

R.R. Hodges Jr., Helium and hydrogen in the lunar atmosphere. J. Geophys. Res. **78**, 8055 (1973). https://doi.org/10.1029/JA078i034p08055

R.R. Hodges Jr., Formation of the lunar atmosphere. Moon **14**, 139–157 (1975). https://doi.org/10.1007/BF00562980

R.R. Hodges Jr., Release of radiogenic gases from the Moon. Phys. Earth Planet. Inter. **14**, 282–288 (1977). https://doi.org/10.1016/0031-9201(77)90178-9

R.R. Hodges Jr., Monte Carlo simulation of the terrestrial hydrogen exosphere. J. Geophys. Res. **99**, 23229 (1994). https://doi.org/10.1029/94JA02183

M. Holmström, Asymmetries in Mars' exosphere. Implications for X-ray and ENA imaging. Space Sci. Rev. **126**, 435–445 (2006). https://doi.org/10.1007/s11214-006-9036-7

M. Holmström, S. Barabash, E. Kallio, X-ray imaging of the solar wind—Mars interaction. Geophys. Res. Lett. **28**, 1287–1290 (2001). https://doi.org/10.1029/2000GL012381

M. Horanyi, T.E. Cravens, J.H. Waite Jr., The precipitation of energetic heavy ions into the upper atmosphere of Jupiter. J. Geophys. Res. **93**, 7251–7271 (1988). https://doi.org/10.1029/JA093iA07p07251

M. Horányi, Z. Sternovsky, M. Lankton, C. Dumont, S. Gagnard, D. Gathright, E. Grün, D. Hansen, D. James, S. Kempf, B. Lamprecht, R. Srama, J.R. Szalay, G. Wright, The Lunar Dust Experiment (LDEX) onboard the Lunar Atmosphere and Dust Environment Explorer (LADEE) mission. Space Sci. Rev. **185**, 93–113 (2014). https://doi.org/10.1007/s11214-014-0118-7

K. Hosokawa, S. Taguchi, S. Suzuki, M.R. Collier, T.E. Moore, M.F. Thomsen, Estimation of magnetopause motion from low-energy neutral atom emission. J. Geophys. Res. Space Phys. **113**, 10205 (2008). https://doi.org/10.1029/2008JA013124

H.C. Howe Jr., J.H. Binsack, Explorer 33 and 35 plasma observations of magnetosheath flow. J. Geophys. Res. **77**, 3334 (1972). https://doi.org/10.1029/JA077i019p03334

Y.Q. Hu, X.C. Guo, C. Wang, On the ionospheric and reconnection potentials of the Earth: results from global MHD simulations. J. Geophys. Res. Space Phys. **112**, 07215 (2007). https://doi.org/10.1029/2006JA012145

D. Hubert, A. Samsonov, Steady state slow shock inside the Earth's magnetosheath: to be or not to be? 1. The original observations revisited. J. Geophys. Res. Space Phys. **109**, 1217 (2004). https://doi.org/10.1029/2003JA010008

D. Hubert, A. Samsonov, Reply to the comment by P. Song et al. on "Steady state slow shock inside the Earth's magnetosheath: to be or not to be? 1. The original observations revisited" by D. Hubert and A. Samsonov. J. Geophys. Res. Space Phys. **110**, 11211 (2005). https://doi.org/10.1029/2005JA011224

Y. Hui, D.R. Schultz, V.A. Kharchenko, P.C. Stancil, T.E. Cravens, C.M. Lisse, A. Dalgarno, The ion-induced charge-exchange X-ray emission of the Jovian auroras: magnetospheric or solar wind origin? Astrophys. J. Lett. **702**, 158–162 (2009). https://doi.org/10.1088/0004-637X/702/2/L158

Y. Hui, D.R. Schultz, V.A. Kharchenko, A. Bhardwaj, G. Branduardi-Raymont, P.C. Stancil, T.E. Cravens, C.M. Lisse, A. Dalgarno, Comparative analysis and variability of the Jovian X-ray spectra detected by the Chandra and XMM-Newton observatories. J. Geophys. Res. Space Phys. **115**, 07102 (2010a). https://doi.org/10.1029/2009JA014854

Y. Hui, T.E. Cravens, N. Ozak, D.R. Schultz, What can be learned from the absence of auroral X-ray emission from Saturn? J. Geophys. Res. Space Phys. **115**, 10239 (2010b). https://doi.org/10.1029/2010JA015639

B. Hultqvist, M. Øieroset, G. Paschmann, R. Treumann, Magnetospheric Plasma Sources and Losses: Final Report of the ISSI Study Project on Source and Loss Processes of Magnetospheric Plasma. Space Sci. Rev. **88**, 406–468 (1999). https://doi.org/10.1023/A:1017251707826

K.-J. Hwang, M.L. Goldstein, M.M. Kuznetsova, Y. Wang, A.F. Viñas, D.G. Sibeck, The first in situ observation of Kelvin-Helmholtz waves at high-latitude magnetopause during strongly dawnward interplanetary magnetic field conditions. J. Geophys. Res. Space Phys. **117**, 08233 (2012). https://doi.org/10.1029/2011JA017256

W.L. Imhof, K.A. Spear, J.W. Hamilton, B.R. Higgins, M.J. Murphy, J.G. Pronko, R.R. Vondrak, D.L. McKenzie, C.J. Rice, D.J. Gorney, D.A. Roux, R.L. Williams, J.A. Stein, J. Bjordal, J. Stadsnes, K. Njoten, T.J. Rosenberg, L. Lutz, D. Detrick, The Polar Ionospheric X-ray Imaging Experiment (PIXIE). Space Sci. Rev. **71**, 385–408 (1995). https://doi.org/10.1007/BF00751336

F.M. Ipavich, A.B. Galvin, G. Gloeckler, D. Dovestadt, B. Klecker, Solar wind Fe and CNO measurements in high-speed flows. J. Geophys. Res. **91**, 4133–4141 (1986). https://doi.org/10.1029/JA091iA04p04133

K. Ishikawa, Suzaku study of solar wind charge exchange X-ray emission from the Earth's exosphere, PhD thesis, Tokyo Metropolitan University (2013)

K. Ishikawa, Y. Ezoe, T. Ohashi, N. Terada, Y. Futaana, X-ray observation of Mars at solar minimum with Suzaku. Publ. Astron. Soc. Jpn. **63**, 705–712 (2011). https://doi.org/10.1093/pasj/63.sp3.S705

K. Ishikawa, Y. Ezoe, Y. Miyoshi, N. Terada, K. Mitsuda, T. Ohashi, Suzaku observation of strong solar-wind charge-exchange emission from the terrestrial exosphere during a geomagnetic storm. Publ. Astron. Soc. Jpn. **65**, 63 (2013)

R.C. Isler, An overview of charge-exchange spectroscopy as a plasma diagnostic. Plasma Phys. Control. Fusion **36**, 171–208 (1994). https://doi.org/10.1088/0741-3335/36/2/001

B.M. Jakosky, J.M. Grebowsky, J.G. Luhmann, J. Connerney, F. Eparvier, R. Ergun, J. Halekas, D. Larson, P. Mahaffy, J. McFadden, D.F. Mitchell, N. Schneider, R. Zurek, S. Bougher, D. Brain, Y.J. Ma, C. Mazelle, L. Andersson, D. Andrews, D. Baird, D. Baker, J.M. Bell, M. Benna, M. Chaffin, P. Chamberlin, Y.-Y. Chaufray, J. Clarke, G. Collinson, M. Combi, F. Crary, T. Cravens, M. Crismani, S. Curry, D. Curtis, J. Deighan, G. Delory, R. Dewey, G. DiBraccio, C. Dong, Y. Dong, P. Dunn, M. Elrod, S. England, A. Eriksson, J. Espley, S. Evans, X. Fang, M. Fillingim, K. Fortier, C.M. Fowler, J. Fox, H. Gröller, S. Guzewich, T. Hara, Y. Harada, G. Holsclaw, S.K. Jain, R. Jolitz, F. Leblanc, C.O. Lee, Y. Lee, F. Lefevre, R. Lillis, R. Livi, D. Lo, M. Mayyasi, W. McClintock, T. McEnulty, R. Modolo, F. Montmessin, M. Morooka, A. Nagy, K. Olsen, W. Peterson, A. Rahmati, S. Ruhunusiri, C.T. Russell, S. Sakai, J.-A. Sauvaud, K. Seki, M. Steckiewicz, M. Stevens, A.I.F. Stewart, A. Stiepen, S. Stone, V. Tenishev, E. Thiemann, R. Tolson, D. Toublanc, M. Vogt, T. Weber, P. Withers, T. Woods, R. Yelle, MAVEN observations of the response of Mars to an interplanetary coronal mass ejection. Science **350**, 0210 (2015a). https://doi.org/10.1126/science.aad0210

B.M. Jakosky, R.P. Lin, J.M. Grebowsky, J.G. Luhmann, D.F. Mitchell, G. Beutelschies, T. Priser, M. Acuna, L. Andersson, D. Baird, D. Baker, R. Bartlett, M. Benna, S. Bougher, D. Brain, D. Carson, S. Cauffman, P. Chamberlin, J.-Y. Chaufray, O. Cheatom, J. Clarke, J. Connerney, T. Cravens, D. Curtis, G. Delory, S. Demcak, A. DeWolfe, F. Eparvier, R. Ergun, A. Eriksson, J. Espley, X. Fang, D. Folta, J. Fox, C. Gomez-Rosa, S. Habenicht, J. Halekas, G. Holsclaw, M. Houghton, R. Howard, M. Jarosz, N. Jedrich, M. Johnson, W. Kasprzak, M. Kelley, T. King, M. Lankton, D. Larson, F. Leblanc, F. Lefevre, R. Lillis, P. Mahaffy, C. Mazelle, W. McClintock, J. McFadden, D.L. Mitchell, F. Montmessin, J. Morrissey, W. Peterson, W. Possel, J.-A. Sauvaud, N. Schneider, W. Sidney, S. Sparacino, A.I.F. Stewart, R. Tolson, D. Toublanc, C. Waters, T. Woods, R. Yelle, R. Zurek, The Mars Atmosphere and Volatile Evolution (MAVEN) mission. Space Sci. Rev. **195**, 3–48 (2015b). https://doi.org/10.1007/s11214-015-0139-x

R.K. Janev, H. Winter, State-selective electron capture in atom-highly charged ion collisions. Phys. Rep. **117**, 265–387 (1985). https://doi.org/10.1016/0370-1573(85)90118-8

R.K. Janev, D.S. Belic, B.H. Bransden, Total and partial cross sections for electron capture in collisions of hydrogen atoms with fully stripped ions. Phys. Rev. A **28**, 1293–1302 (1983). https://doi.org/10.1103/PhysRevA.28.1293

R.K. Janev, R.A. Phaneuf, H.T. Hunter, Recommended cross sections for electron capture and ionization in collisions of C^{q+} and O^{q+} ions with H, He and H_2. At. Data Nucl. Data Tables **40**, 249 (1988). https://doi.org/10.1016/0092-640X(88)90008-3

F. Jansen, D. Lumb, B. Altieri, J. Clavel, M. Ehle, C. Erd, C. Gabriel, M. Guainazzi, P. Gondoin, R. Much, R. Munoz, M. Santos, N. Schartel, D. Texier, G. Vacanti, XMM-Newton observatory. I. The spacecraft and operations. Astron. Astrophys. **365**, 1–6 (2001). https://doi.org/10.1051/0004-6361:20000036

K. Jelínek, Z. Němeček, J. Šafránková, Simultaneous observations of the bow shock and magnetopause motions, in *WDS 2006—Proceedings of Contributed Papers: Part II—Physics of Plasmas and Ionized Media*, ed. by J. Šafránková, J. Pavløu (2006), pp. 14–20

K. Jelínek, Z. Němeček, J. Šafránková, J.-H. Shue, A.V. Suvorova, D.G. Sibeck, Thin magnetosheath as a consequence of the magnetopause deformation: THEMIS observations. J. Geophys. Res. Space Phys. **115**, 10203 (2010). https://doi.org/10.1029/2010JA015345

S.R. Jelinsky, O.H. Siegmund, J.A. Mir, Progress in soft x-ray and UV photocathodes, in *EUV, X-Ray, and Gamma-Ray Instrumentation for Astronomy VII*, ed. by O.H. Siegmund, M.A. Gummin. Proc. SPIE., vol. 2808 (1996), pp. 617–625. https://doi.org/10.1117/12.256036

M. Jeřáb, Z. Němeček, J. Šafránková, K. Jelínek, J. Měrka, Improved bow shock model with dependence on the IMF strength. Planet. Space Sci. **53**, 85–93 (2005). https://doi.org/10.1016/j.pss.2004.09.032

J.R. Kan, A theory of patchy and intermittent reconnections for magnetospheric flux transfer events. J. Geophys. Res. **93**, 5613–5623 (1988). https://doi.org/10.1029/JA093iA06p05613

R.L. Kaufmann, A. Konradi, Explorer 12 magnetopause observations: large-scale nonuniform motion. J. Geophys. Res. **74**, 3609 (1969). https://doi.org/10.1029/JA074i014p03609

S. Kavosi, J. Raeder, Ubiquity of Kelvin-Helmholtz waves at Earth's magnetopause. Nat. Commun. **6**, 7019 (2015). https://doi.org/10.1038/ncomms8019

H. Kawano, C.T. Russell, Cause of postterminator flux transfer events. J. Geophys. Res. **102**, 27029–27038 (1997). https://doi.org/10.1029/97JA02139

V. Kharchenko, A. Dalgarno, Spectra of cometary X rays induced by solar wind ions. J. Geophys. Res. **105**, 18351–18360 (2000). https://doi.org/10.1029/1999JA000203

V. Kharchenko, A. Dalgarno, Variability of cometary X-ray emission induced by solar wind ions. Astrophys. J. Lett. **554**, 99–102 (2001). https://doi.org/10.1086/320929

W.C. Knudsen, A.J. Kliore, R.C. Whitten, Solar cycle changes in the ionization sources of the nightside Venus ionosphere. J. Geophys. Res. **92**, 13391–13398 (1987). https://doi.org/10.1029/JA092iA12p13391

W. Köhnlein, Radial dependence of solar wind parameters in the ecliptic (1.1 R$_\odot$–61 AU). Sol. Phys. **169**, 209–213 (1996). https://doi.org/10.1007/BF00153841

G.I. Korotova, D.G. Sibeck, N. Omidi, V. Angelopoulos, THEMIS observations of unusual bow shock motion attending a transient magnetospheric event. J. Geophys. Res. Space Phys. **117**, 12207 (2012). https://doi.org/10.1029/2012JA017510

D. Koutroumpa, Update on modeling and data analysis of heliospheric solar wind charge exchange X-ray emission. Astron. Nachr. **333**, 341 (2012). https://doi.org/10.1002/asna.201211666

D. Koutroumpa, R. Lallement, V. Kharchenko, A. Dalgarno, R. Pepino, V. Izmodenov, E. Quémerais, Charge-transfer induced EUV and soft X-ray emissions in the heliosphere. Astron. Astrophys. **460**, 289–300 (2006). https://doi.org/10.1051/0004-6361:20065250

D. Koutroumpa, F. Acero, R. Lallement, J. Ballet, V. Kharchenko, OVII and OVIII line emission in the diffuse soft X-ray background: heliospheric and galactic contributions. Astron. Astrophys. **475**, 901–914 (2007). https://doi.org/10.1051/0004-6361:20078271

D. Koutroumpa, M.R. Collier, K.D. Kuntz, R. Lallement, S.L. Snowden, Solar wind charge exchange emission from the helium focusing cone: model to data comparison. Astrophys. J. **697**, 1214–1225 (2009a). https://doi.org/10.1088/0004-637X/697/2/1214

D. Koutroumpa, R. Lallement, J.C. Raymond, V. Kharchenko, The solar wind charge-transfer X-ray emission in the 1/4 keV energy range: inferences on local bubble hot gas at low Z. Astrophys. J. **696**, 1517–1525 (2009b). https://doi.org/10.1088/0004-637X/696/2/1517

D. Koutroumpa, R. Modolo, G. Chanteur, J.-Y. Chaufray, V. Kharchenko, R. Lallement, Solar wind charge exchange X-ray emission from Mars. Model and data comparison. Astron. Astrophys. **545**, 153 (2012). https://doi.org/10.1051/0004-6361/201219720

K. Koyama, H. Tsunemi, T. Dotani, M.W. Bautz, K. Hayashida, T.G. Tsuru, H. Matsumoto, Y. Ogawara, G.R. Ricker, J. Doty, S.E. Kissel, R. Foster, H. Nakajima, H. Yamaguchi, H. Mori, M. Sakano, K. Hamaguchi, M. Nishiuchi, E. Miyata, K. Torii, M. Namiki, S. Katsuda, D. Matsuura, T. Miyauchi, N. Anabuki, N. Tawa, M. Ozaki, H. Murakami, Y. Maeda, Y. Ichikawa, G.Y. Prigozhin, E.A. Boughan, B. Lamarr, E.D. Miller, B.E. Burke, J.A. Gregory, A. Pillsbury, A. Bamba, J.S. Hiraga, A. Senda, H. Katayama, S. Kitamoto, M. Tsujimoto, T. Kohmura, Y. Tsuboi, H. Awaki, X-ray Imaging Spectrometer (XIS) on board Suzaku. Publ. Astron. Soc. Jpn. **59**, 23–33 (2007). https://doi.org/10.1093/pasj/59.sp1.S23

V. Krasnopolsky, NOTE: on the deuterium abundance on Mars and some related problems. Icarus **148**, 597–602 (2000). https://doi.org/10.1006/icar.2000.6534

V.A. Krasnopolsky, J.B. Greenwood, P.C. Stancil, X-ray and extreme ultraviolet emissions from comets. Space Sci. Rev. **113**, 271–374 (2004). https://doi.org/10.1023/B:SPAC.0000046754.75560.80

K.D. Kuntz, S.L. Snowden, Deconstructing the spectrum of the soft X-ray background. Astrophys. J. **543**, 195–215 (2000). https://doi.org/10.1086/317071

K.D. Kuntz, S.L. Snowden, The EPIC-MOS particle-induced background spectra. Astron. Astrophys. **478**, 575–596 (2008). https://doi.org/10.1051/0004-6361:20077912

K.D. Kuntz, Y.M. Collado-Vega, M.R. Collier, H.K. Connor, T.E. Cravens, D. Koutroumpa, F.S. Porter, I.P. Robertson, D.G. Sibeck, S.L. Snowden, N.E. Thomas, B.M. Walsh, The solar wind charge-exchange production factor for hydrogen. Astrophys. J. **808**, 143 (2015). https://doi.org/10.1088/0004-637X/808/2/143

T.V. Laitinen, M. Palmroth, T.I. Pulkkinen, P. Janhunen, H.E.J. Koskinen, Continuous reconnection line and pressure-dependent energy conversion on the magnetopause in a global MHD model. J. Geophys. Res. Space Phys. **112**, 11201 (2007). https://doi.org/10.1029/2007JA012352

R. Lallement, The heliospheric soft X-ray emission pattern during the ROSAT survey: inferences on Local Bubble hot gas. Astron. Astrophys. **418**, 143–150 (2004). https://doi.org/10.1051/0004-6361:20040059

R. Lallement, J.L. Bertaux, F. Dalaudier, Interplanetary Lyman-alpha spectral profiles and intensities for both repulsive and attractive solar force fields predicted absorption pattern by a hydrogen cell. Astron. Astrophys. **150**, 21–32 (1985a).

R. Lallement, J.L. Bertaux, V.G. Kurt, Solar wind decrease at high heliographic latitudes detected from Prognoz interplanetary Lyman alpha mapping. J. Geophys. Res. **90**, 1413–1423 (1985b). https://doi.org/10.1029/JA090iA02p01413

R. Lallement, J.C. Raymond, J. Vallerga, M. Lemoine, F. Dalaudier, J.L. Bertaux, Modeling the interstellar-interplanetary helium 58.4 nm resonance glow: towards a reconciliation with particle measurements. Astron. Astrophys. **426**, 875–884 (2004a). https://doi.org/10.1051/0004-6361:20035929

R. Lallement, J.C. Raymond, J.-L. Bertaux, E. Quémerais, Y.-K. Ko, M. Uzzo, D. McMullin, D. Rucinski, Solar cycle dependence of the helium focusing cone from SOHO/UVCS observations.

 Springer

Electron impact rates and associated pickup ions. Astron. Astrophys. **426**, 867–874 (2004b). https://doi.org/10.1051/0004-6361:200400028

R. Lallement, E. Quémerais, J.L. Bertaux, S. Ferron, D. Koutroumpa, R. Pellinen, Deflection of the interstellar neutral hydrogen flow across the heliospheric interface. Science **307**, 1447–1449 (2005). https://doi.org/10.1126/science.1107953

H. Lammer, H.I.M. Lichtenegger, H.K. Biernat, N.V. Erkaev, I.L. Arshukova, C. Kolb, H. Gunell, A. Lukyanov, M. Holmstrom, S. Barabash, T.L. Zhang, W. Baumjohann, Loss of hydrogen and oxygen from the upper atmosphere of Venus. Planet. Space Sci. **54**, 1445–1456 (2006). https://doi.org/10.1016/j.pss.2006.04.022

G. Lapenta, D. Krauss-Varban, H. Karimabadi, J.D. Huba, L.I. Rudakov, P. Ricci, Kinetic simulations of x-line expansion in 3D reconnection. Geophys. Res. Lett. **33**, 10102 (2006). https://doi.org/10.1029/2005GL025124

B. Lavraud, T. Phan, M. Dunlop, M. Taylor, P. Cargill, J. Bosqued, I. Dandouras, H. Rème, J. Sauvaud, C. Escoubet, A. Balogh, A. Fazakerley, The exterior cusp and its boundary with the magnetosheath: cluster multi-event analysis. Ann. Geophys. **22**, 3039–3054 (2004). https://doi.org/10.5194/angeo-22-3039-2004

G. Le, C.T. Russell, H. Kuo, Flux transfer events—spontaneous or driven? Geophys. Res. Lett. **20**, 791–794 (1993). https://doi.org/10.1029/93GL00850

L.C. Lee, M. Yan, J.G. Hawkins, A study of slow-mode structures in the dayside magnetosheath. Geophys. Res. Lett. **18**, 381–384 (1991). https://doi.org/10.1029/90GL02787

W.Y. Li, X.C. Guo, C. Wang, Spatial distribution of Kelvin-Helmholtz instability at low-latitude boundary layer under different solar wind speed conditions. J. Geophys. Res. Space Phys. **117**, 08230 (2012). https://doi.org/10.1029/2012JA017780

R.J. Lillis, D.A. Brain, S.W. Bougher, F. Leblanc, J.G. Luhmann, B.M. Jakosky, R. Modolo, J. Fox, J. Deighan, X. Fang, Y.C. Wang, Y. Lee, C. Dong, Y. Ma, T. Cravens, L. Andersson, S.M. Curry, N. Schneider, M. Combi, I. Stewart, J. Clarke, J. Grebowsky, D.L. Mitchell, R. Yelle, A.F. Nagy, D. Baker, R.P. Lin, Characterizing atmospheric escape from Mars today and through time, with MAVEN. Space Sci. Rev. **195**, 357–422 (2015). https://doi.org/10.1007/s11214-015-0165-8

R.L. Lin, X.X. Zhang, S.Q. Liu, Y.L. Wang, J.C. Gong, A three-dimensional asymmetric magnetopause model. J. Geophys. Res. Space Phys. **115**, 4207 (2010). https://doi.org/10.1029/2009JA014235

C.M. Lisse, K. Dennerl, J. Englhauser, M. Harden, F.E. Marshall, M.J. Mumma, R. Petre, J.P. Pye, M.J. Ricketts, J. Schmitt, J. Trumper, R.G. West, Discovery of X-ray and extreme ultraviolet emission from comet C/Hyakutake 1996 B2. Science **274**, 205–209 (1996). https://doi.org/10.1126/science.274.5285.205

C.M. Lisse, D.J. Christian, K. Dennerl, K.J. Meech, R. Petre, H.A. Weaver, S.J. Wolk, Charge exchange-induced X-ray emission from comet C/1999 S4 (LINEAR). Science **292**, 1343–1348 (2001). https://doi.org/10.1126/science.292.5520.1343

C.M. Lisse, D.J. Christian, K. Dennerl, S.J. Wolk, D. Bodewits, R. Hoekstra, M.R. Combi, T. Mäkinen, M. Dryer, C.D. Fry, H. Weaver, Chandra observations of comet 2P/Encke 2003: first detection of a collisionally thin, fast solar wind charge exchange system. Astrophys. J. **635**, 1329–1347 (2005). https://doi.org/10.1086/497570

C.M. Lisse, K. Dennerl, D.J. Christian, S.J. Wolk, D. Bodewits, T.H. Zurbuchen, K.C. Hansen, R. Hoekstra, M. Combi, C.D. Fry, M. Dryer, T. Mäkinen, W. Sun, Chandra observations of comet 9P/Tempel 1 during the Deep Impact campaign. Icarus **191**, 295–309 (2007). https://doi.org/10.1016/j.icarus.2007.03.038

C.M. Lisse, R.L. McNutt, S.J. Wolk, F. Bagenal, S.A. Stern, G.R. Gladstone, T.E. Cravens, M.E. Hill, P. Kollmann, H.A. Weaver, D.F. Strobel, H.A. Elliott, D.J. McComas, R.P. Binzel, B.T. Snios, A. Bhardwaj, A. Chutjian, L.A. Young, C.B. Olkin, K.A. Ennico, The puzzling detection of x-rays from Pluto by Chandra. Icarus **287**, 103–109 (2017). https://doi.org/10.1016/j.icarus.2016.07.008

W. Liu, M. Chiao, M.R. Collier, T. Cravens, M. Galeazzi, D. Koutroumpa, K.D. Kuntz, R. Lallement, S.T. Lepri, D. McCammon, K. Morgan, F.S. Porter, S.L. Snowden, N.E. Thomas, Y. Uprety, E. Ursino, B.M. Walsh, The structure of the local hot bubble. Astrophys. J. **834**, 33 (2017). https://doi.org/10.3847/1538-4357/834/1/33

M. Lockwood, M.N. Wild, On the quasi-periodic nature of magnetopause flux transfer events. J. Geophys. Res. **98**, 5935–5940 (1993). https://doi.org/10.1029/92JA02375

M. Lockwood, P.E. Sandholt, S.W.H. Cowley, T. Oguti, Interplanetary magnetic field control of dayside auroral activity and the transfer of momentum across the dayside magnetopause. Planet. Space Sci. **37**, 1347–1365 (1989). https://doi.org/10.1016/0032-0633(89)90106-2

M. Lockwood, S.W.H. Cowley, P.E. Sandholt, R.P. Lepping, The ionospheric signatures of flux transfer events and solar wind dynamic pressure changes. J. Geophys. Res. **95**, 17113–17135 (1990). https://doi.org/10.1029/JA095iA10p17113

M. Lockwood, S.W.H. Cowley, M.F. Smith, R.P. Rijnbeek, R.C. Elphic, The contribution of flux transfer events to convection. Geophys. Res. Lett. **22**, 1185–1188 (1995). https://doi.org/10.1029/95GL01008

M. Longmore, S.J. Schwartz, J. Geach, B.M.A. Cooling, I. Dandouras, E.A. Lucek, A.N. Fazakerley, Dawn-dusk asymmetries and sub-Alfvénic flow in the high and low latitude magnetosheath. Ann. Geophys. **23**, 3351–3364 (2005). https://doi.org/10.5194/angeo-23-3351-2005

J.Y. Lu, Z.-Q. Liu, K. Kabin, M.X. Zhao, D.D. Liu, Q. Zhou, Y. Xiao, Three dimensional shape of the magnetopause: global MHD results. J. Geophys. Res. Space Phys. **116**, 09237 (2011). https://doi.org/10.1029/2010JA016418

J.G. Luhmann, W.T. Kasprzak, C.T. Russell, Space weather at Venus and its potential consequences for atmosphere evolution. J. Geophys. Res., Planets **112**, 4–10 (2007). https://doi.org/10.1029/2006JE002820

D.H. Lumb, R.S. Warwick, M. Page, A. De Luca, X-ray background measurements with XMM-Newton EPIC. Astron. Astrophys. **389**, 93–105 (2002). https://doi.org/10.1051/0004-6361:20020531

R. Lundin, Ion acceleration and outflow from Mars and Venus: an overview. Space Sci. Rev. **162**, 309–334 (2011). https://doi.org/10.1007/s11214-011-9811-y

R. Lundin, S. Barabash, M. Holmström, H. Nilsson, Y. Futaana, R. Ramstad, M. Yamauchi, E. Dubinin, M. Fraenz, Solar cycle effects on the ion escape from Mars. Geophys. Res. Lett. **40**, 6028–6032 (2013). https://doi.org/10.1002/2013GL058154

E.F. Lyon, H.S. Bridge, J.H. Binsack, Explorer 35 plasma measurements in the vicinity of the Moon. J. Geophys. Res. **72**, 6113 (1967). https://doi.org/10.1029/JZ072i023p06113

J.G. Lyon, MHD simulations of the magnetosheath. Adv. Space Res. **14**, 21–28 (1994). https://doi.org/10.1016/0273-1177(94)90043-4

Y. Ma, A.F. Nagy, K.C. Hansen, D.L. Dezeeuw, T.I. Gombosi, K.G. Powell, Three-dimensional multispecies MHD studies of the solar wind interaction with Mars in the presence of crustal fields. J. Geophys. Res. Space Phys. **107**, 1282 (2002). https://doi.org/10.1029/2002JA009293

K. Maezawa, Magnetotail boundary motion associated with geomagnetic substorms. J. Geophys. Res. **80**, 3543–3548 (1975). https://doi.org/10.1029/JA080i025p03543

P.R. Mahaffy, R.R. Hodges, M. Benna, T.T. King, R. Arvey, M. Barciniak, M. Bendt, D. Carigan, T. Errigo, et al., The Neutral Mass Spectrometer on the Lunar Atmosphere and Dust Environment Explorer mission. Space Sci. Rev. **185**, 27–61 (2014). https://doi.org/10.1007/s11214-014-0043-9

I.P. Maltsev, V.B. Lyatsky, Field-aligned currents and erosion of the dayside magnetosphere. Planet. Space Sci. **23**, 1257–1260 (1975). https://doi.org/10.1016/0032-0633(75)90149-X

R. Mann, H.F. Beyer, F. Folkmann, Selective electron capture: a dominant production process for few-electron states of light target atoms after heavy-ion impact. Phys. Rev. Lett. **46**, 646–650 (1981). https://doi.org/10.1103/PhysRevLett.46.646, https://link.aps.org/doi/10.1103/PhysRevLett.46.646

A. Martindale, J.F. Pearson, G.W. Fraser, J.D. Carpenter, R. Willingale, T. Stevenson, C. Whitford, F. Giannini, R. Fairbend, J. Seguy, E. Sclater, I. Delgado, M. Kaipiainen, S. Nenonen, T. Pilvi, E. Schyns, C. Bulloch, C. Sawyers, K. Muinonen, The Mercury Imaging X-ray Spectrometer: optics design and characterisation, in *Instruments and Methods for Astrobiology and Planetary Missions XII*. Proc. SPIE., vol. 7441 (2009), p. 744117. https://doi.org/10.1117/12.826083

C. Martinecz, M. Fränz, J. Woch, N. Krupp, E. Roussos, E. Dubinin, U. Motschmann, S. Barabash, R. Lundin, M. Holmström, H. Andersson, M. Yamauchi, A. Grigoriev, Y. Futaana, K. Brinkfeldt, H. Gunell, R.A. Frahm, J.D. Winningham, J.R. Sharber, J. Scherrer, A.J. Coates, D.R. Linder, D.O. Kataria, E. Kallio, T. Sales, W. Schmidt, P. Riihela, H.E.J. Koskinen, J.U. Kozyra, J. Luhmann, C.T. Russell, E.C. Roelof, P. Brandt, C.C. Curtis, K.C. Hsieh, B.R. Sandel, M. Grande, J.-A. Sauvaud, A. Fedorov, J.-J. Thocaven, C. Mazelle, S. McKenna-Lawler, O. Orsini, R. Cerulli-Irelli, M. Maggi, A. Mura, A. Milillo, P. Wurz, A. Galli, P. Bochsler, K. Asamura, K. Szego, W. Baumjohann, T.L. Zhang, H. Lammer, Location of the bow shock and ion composition boundaries at Venus—initial determinations from Venus Express ASPERA-4. Planet. Space Sci. **56**, 780–784 (2008). https://doi.org/10.1016/j.pss.2007.07.007

C. Martinecz, A. Boesswetter, M. Fränz, E. Roussos, J. Woch, N. Krupp, E. Dubinin, U. Motschmann, S. Wiehle, S. Simon, S. Barabash, R. Lundin, T.L. Zhang, H. Lammer, H. Lichtenegger, Y. Kulikov, Plasma environment of Venus: comparison of Venus Express ASPERA-4 measurements with 3-D hybrid simulations. J. Geophys. Res. Space Phys. **114**, E00B30 (2009). https://doi.org/10.1029/2008JE003174

R.J. Mawhorter, A. Chutjian, T.E. Cravens, N. Djurić, S. Hossain, C.M. Lisse, J.A. Macaskill, S.J. Smith, J. Simcic, I.D. Williams, Absolute single and multiple charge exchange cross sections for highly charged C, O, and Ne ions on H_2O, CO, and CO^2. Phys. Rev. A **75**(3), 032704 (2007). https://doi.org/10.1103/PhysRevA.75.032704

N. Maynard, W. Burke, J. Scudder, D. Ober, G. Siscoe, W. White, K. Siebert, D. Weimer, G. Erickson, J. Schoendorf, M. Heinemann, Observed and simulated depletion layers with southward IMF. Ann. Geophys. **22**, 2151–2169 (2004). https://doi.org/10.5194/angeo-22-2151-2004

D. McCammon, W.T. Sanders, The soft X-ray background and its origins. Annu. Rev. Astron. Astrophys. **28**, 657–688 (1990). https://doi.org/10.1146/annurev.aa.28.090190.003301

D.J. McComas, S.J. Bame, B.L. Barraclough, J.R. Donart, R.C. Elphic, J.T. Gosling, M.B. Moldwin, K.R. Moore, M.F. Thomsen, Magnetospheric plasma analyzer—initial three-spacecraft observations from geosynchronous orbit. J. Geophys. Res. **98**, 13453 (1993). https://doi.org/10.1029/93JA00726

D.J. McComas, R.C. Elphic, M.B. Moldwin, M.F. Thomsen, Plasma observations of magnetopause crossing at geosynchronous orbit. J. Geophys. Res. **99**, 21249 (1994). https://doi.org/10.1029/94JA01094

D.J. McComas, S.J. Bame, P. Barker, W.C. Feldman, J.L. Phillips, P. Riley, J.W. Griffee, Solar Wind Electron Proton Alpha Monitor (SWEPAM) for the Advanced Composition Explorer. Space Sci. Rev. **86**, 563–612 (1998). https://doi.org/10.1023/A:1005040232597

D.J. McComas, F. Allegrini, P. Bochsler, M. Bzowski, M. Collier, H. Fahr, H. Fichtner, P. Frisch, H.O. Funsten, S.A. Fuselier, G. Gloeckler, M. Gruntman, V. Izmodenov, P. Knappenberger, M. Lee, S. Livi, D. Mitchell, E. Möbius, T. Moore, S. Pope, D. Reisenfeld, E. Roelof, J. Scherrer, N. Schwadron, R. Tyler, M. Wieser, M. Witte, P. Wurz, G. Zank, IBEX—Interstellar Boundary Explorer. Space Sci. Rev. **146**, 11–33 (2009). https://doi.org/10.1007/s11214-009-9499-4

S.B. Mende, H.U. Frey, T.J. Immel, J.-C. Gerard, B. Hubert, S.A. Fuselier, Global imaging of proton and electron aurorae in the far ultraviolet. Space Sci. Rev. **109**, 211–254 (2003). https://doi.org/10.1023/B:SPAC.0000007520.23689.08

D.A. Mendis, A postencounter view of comets. Annu. Rev. Astron. Astrophys. **26**, 11–49 (1988). https://doi.org/10.1146/annurev.aa.26.090188.000303

D.A. Mendis, M. Horányi, The global morphology of the solar wind interaction with comet Churyumov-Gerasimenko. Astrophys. J. **794**, 14 (2014). https://doi.org/10.1088/0004-637X/794/1/14

J. Merka, J. Safránková, Z. Nemecek, Cusp-like plasma in high altitudes: a statistical study of the width and location of the cusp from Magion-4. Ann. Geophys. **20**, 311–320 (2002). https://doi.org/10.5194/angeo-20-311-2002

J. Merka, A. Szabo, T.W. Narock, J.H. King, K.I. Paularena, J.D. Richardson, A comparison of IMP 8 observed bow shock positions with model predictions. J. Geophys. Res. Space Phys. **108**, 1077 (2003a). https://doi.org/10.1029/2002JA009384

J. Merka, A. Szabo, J. Šafránková, Z. Němeček, Earth's bow shock and magnetopause in the case of a field-aligned upstream flow: observation and model comparison. J. Geophys. Res. Space Phys. **108**, 1269 (2003b). https://doi.org/10.1029/2002JA009697

J. Merka, A. Szabo, J.A. Slavin, M. Peredo, Three-dimensional position and shape of the bow shock and their variation with upstream Mach numbers and interplanetary magnetic field orientation. J. Geophys. Res. Space Phys. **110**, 04202 (2005). https://doi.org/10.1029/2004JA010944

A.E. Metzger, D.A. Gilman, J.L. Luthey, K.C. Hurley, H.W. Schnopper, F.D. Seward, J.D. Sullivan, The detection of X rays from Jupiter. J. Geophys. Res. **88**, 7731–7741 (1983). https://doi.org/10.1029/JA088iA10p07731

D.G. Mitchell, E.C. Roelof, S.J. Bame, Solar wind iron abundance variations at speeds greater than 600 km/s, 1972–1976. J. Geophys. Res. **88**, 9059–9068 (1983). https://doi.org/10.1029/JA088iA11p09059

I.G. Mitrofanov, A.B. Sanin, W.V. Boynton, G. Chin, J.B. Garvin, D. Golovin, L.G. Evans, K. Harshman, A.S. Kozyrev, M.L. Litvak, A. Malakhov, E. Mazarico, T. McClanahan, G. Milikh, M. Mokrousov, G. Nandikotkur, G.A. Neumann, I. Nuzhdin, R. Sagdeev, V. Shevchenko, V. Shvetsov, D.E. Smith, R. Starr, V.I. Tretyakov, J. Trombka, D. Usikov, A. Varenikov, A. Vostrukhin, M.T. Zuber, Hydrogen mapping of the lunar South Pole using the LRO Neutron Detector experiment LEND. Science **330**, 483 (2010). https://doi.org/10.1126/science.1185696

I. Mitsuishi, Y. Ezoe, T. Ogawa, M. Sato, K. Nakamura, M. Numazawa, K. Takeuchi, T. Ohashi, K. Ishikawa, K. Mitsuda, Ray-tracing simulations for the ultra-lightweight X-ray optics toward a future Jupiter exploration mission. Adv. Space Res. **57**, 320–328 (2016). https://doi.org/10.1016/j.asr.2015.08.022

A. Miura, Anomalous transport by magnetohydrodynamic Kelvin-Helmholtz instabilities in the solar wind-magnetosphere interaction. J. Geophys. Res. **89**, 801–818 (1984). https://doi.org/10.1029/JA089iA02p00801

E. Möbius, P. Bochsler, M. Bzowski, D. Heirtzler, M.A. Kubiak, H. Kucharek, M.A. Lee, T. Leonard, N.A. Schwadron, X. Wu, S.A. Fuselier, G. Crew, D.J. McComas, L. Petersen, L. Saul, D. Valovcin, R. Vanderspek, P. Wurz, Interstellar gas flow parameters derived from interstellar boundary explorer-Lo observations in 2009 and 2010: analytical analysis. Astrophys. J. Suppl. Ser. **198**, 11 (2012). https://doi.org/10.1088/0067-0049/198/2/11

R. Modolo, G.M. Chanteur, E. Dubinin, A.P. Matthews, Simulated solar wind plasma interaction with the Martian exosphere: influence of the solar EUV flux on the bow shock and the magnetic pile-up boundary. Ann. Geophys. **24**, 3403–3410 (2006). https://doi.org/10.5194/angeo-24-3403-2006

M.B. Moldwin, Outer plasmaspheric plasma properties: what we know from satellite data. Space Sci. Rev. **80**, 181–198 (1997). https://doi.org/10.1023/A:1004921903897

K.R. Moore, V.A. Thomas, D.J. McComas, Global hybrid simulation of the solar wind interaction with the dayside of Venus. J. Geophys. Res. **96**, 7779–7791 (1991). https://doi.org/10.1029/91JA00013

T. Moretto, D.G. Sibeck, B. Lavraud, K.J. Trattner, H. Rème, A. Balogh, Flux pile-up and plasma depletion at the high latitude dayside magnetopause during southward interplanetary magnetic field: a cluster event study. Ann. Geophys. **23**, 2259–2264 (2005). https://doi.org/10.5194/angeo-23-2259-2005

D.D. Morgan, C. Diéval, D.A. Gurnett, F. Duru, E.M. Dubinin, M. Fränz, D.J. Andrews, H.J. Opgenoorth, D. Uluşen, I. Mitrofanov, J.J. Plaut, Effects of a strong ICME on the Martian ionosphere as detected by Mars Express and Mars Odyssey. J. Geophys. Res. Space Phys. **119**, 5891–5908 (2014). https://doi.org/10.1002/2013JA019522

U.V. Möstl, N.V. Erkaev, M. Zellinger, H. Lammer, H. Gröller, H.K. Biernat, D. Korovinskiy, The Kelvin-Helmholtz instability at Venus: what is the unstable boundary? Icarus **216**, 476–484 (2011). https://doi.org/10.1016/j.icarus.2011.09.012

S. Mühlbachler, C.J. Farrugia, J. Raeder, H.K. Biernat, R.B. Torbert, A statistical investigation of dayside magnetosphere erosion showing saturation of response. J. Geophys. Res. Space Phys. **110**, 11207 (2005). https://doi.org/10.1029/2005JA011177

J.S. Murphree, R.A. King, T. Payne, K. Smith, D. Reid, J. Adema, B. Gordon, R. Wlochowicz, The Freja ultraviolet imager. Space Sci. Rev. **70**, 421–446 (1994). https://doi.org/10.1007/BF00756880

J.-L. Mutz, O. Bonnet, R. Fairbend, E. Schyns, J. Seguy, Micro-pore optics: from planetary x-rays to industrial market, in *Quantum Sensing and Nanophotonic Devices IV*. Proc. SPIE., vol. 6479 (2007), p. 64790. https://doi.org/10.1117/12.699576

A.F. Nagy, D. Winterhalter, K. Sauer, T.E. Cravens, S. Brecht, C. Mazelle, D. Crider, E. Kallio, A. Zakharov, E. Dubinin, M. Verigin, G. Kotova, W.I. Axford, C. Bertucci, J.G. Trotignon, The plasma environment of Mars. Space Sci. Rev. **111**, 33–114 (2004). https://doi.org/10.1023/B:SPAC.0000032718.47512.92

M. Nakamura, I. Yoshikawa, A. Yamazaki, K. Shiomi, Y. Takizawa, M. Hirahara, K. Yamashita, Y. Saito, W. Miyake, Terrestrial plasmaspheric imaging by an extreme ultraviolet scanner on planet-B. Geophys. Res. Lett. **27**, 141–144 (2000). https://doi.org/10.1029/1999GL010732

T.K.M. Nakamura, M. Fujimoto, A. Otto, Magnetic reconnection induced by weak Kelvin-Helmholtz instability and the formation of the low-latitude boundary layer. Geophys. Res. Lett. **33**, 14106 (2006). https://doi.org/10.1029/2006GL026318

M. Neugebauer, T.E. Cravens, C.M. Lisse, F.M. Ipavich, D. Christian, R. von Steiger, P. Bochsler, P.D. Shah, T.P. Armstrong, The relation of temporal variations of soft X-ray emission from comet Hyakutake to variations of ion fluxes in the solar wind. J. Geophys. Res. **105**, 20949–20956 (2000). https://doi.org/10.1029/1999JA000299

P.T. Newell, C.-I. Meng, Hemispherical asymmetry in cusp precipitation near solstices. J. Geophys. Res. **93**, 2643–2648 (1988). https://doi.org/10.1029/JA093iA04p02643

P.T. Newell, C.-I. Meng, Dipole tilt angle effects on the latitude of the cusp and cleft/low-latitude boundary layer. J. Geophys. Res. **94**, 6949–6953 (1989). https://doi.org/10.1029/JA094iA06p06949

P.T. Newell, C.-I. Meng, Ion acceleration at the equatorward edge of the cusp—low altitude observations of patchy merging. Geophys. Res. Lett. **18**, 1829–1832 (1991). https://doi.org/10.1029/91GL02088

P.T. Newell, C.-I. Meng, Mapping the dayside ionosphere to the magnetosphere according to particle precipitation characteristics. Geophys. Res. Lett. **19**, 609–612 (1992). https://doi.org/10.1029/92GL00404

P.T. Newell, C.-I. Meng, Ionospheric projections of magnetospheric regions under low and high solar wind pressure conditions. J. Geophys. Res. **99**, 273–286 (1994). https://doi.org/10.1029/93JA02273

P.T. Newell, C.-I. Meng, D.G. Sibeck, R. Lepping, Some low-altitude cusp dependencies on the interplanetary magnetic field. J. Geophys. Res. **94**, 8921–8927 (1989). https://doi.org/10.1029/JA094iA07p08921

A. Nishida, Can random reconnection on the magnetopause produce the low latitude boundary layer? Geophys. Res. Lett. **16**, 227–230 (1989). https://doi.org/10.1029/GL016i003p00227

M.N. Nishino, M. Fujimoto, T.-D. Phan, T. Mukai, Y. Saito, M.M. Kuznetsova, L. Rastätter, Anomalous flow deflection at Earth's low-Alfvén-Mach-number bow shock. Phys. Rev. Lett. **101**(6), 065003 (2008). https://doi.org/10.1103/PhysRevLett.101.065003

Z. Němeček, J. Šafránková, IMF control of the high-altitude cusp dynamics. Adv. Space Res. **41**, 92–102 (2008). https://doi.org/10.1016/j.asr.2007.07.038

Z. Němeček, J. Šafránková, G.N. Zastenker, P. Pišoft, K.I. Paularena, Spatial distribution of the magnetosheath ion flux. Adv. Space Res. **30**, 2751–2756 (2002). https://doi.org/10.1016/S0273-1177(02)80402-1

Z. Němeček, J. Šimunek, J. Šafránková, L. Prech, Spatial and temporal variations of the high-altitude cusp precipitation. Ann. Geophys. **22**, 2441–2450 (2004). https://doi.org/10.5194/angeo-22-2441-2004

K. Nykyri, Impact of MHD shock physics on magnetosheath asymmetry and Kelvin-Helmholtz instability. J. Geophys. Res. Space Phys. **118**, 5068–5081 (2013). https://doi.org/10.1002/jgra.50499

D.M. Ober, M.F. Thomsen, N.C. Maynard, Observations of bow shock and magnetopause crossings from geosynchronous orbit on 31 March 2001. J. Geophys. Res. Space Phys. **107**, 1206 (2002). https://doi.org/10.1029/2001JA000284

D.M. Ober, N.C. Maynard, W.J. Burke, G.R. Wilson, K.D. Siebert, "Shoulders" on the high-latitude magnetopause: Polar/GOES observations. J. Geophys. Res. Space Phys. **111**, 10213 (2006). https://doi.org/10.1029/2006JA011799

D. Odstrcil, Modeling 3-D solar wind structure. Adv. Space Res. **32**, 497–506 (2003). https://doi.org/10.1016/S0273-1177(03)00332-6

T. Ogawa, Y. Ezoe, T. Kakiuchi, M. Ikuta, M. Sato, K. Nakamura, M. Numazawa, K. Takeuchi, M. Terada, T. Ohashi, I. Mitsuishi, K. Ishikawa, K. Mitsuda, K. Morishita, K. Nakajima, First x-ray imaging with a micromachined Wolter type-I telescope. Microsyst. Technol. **23**(4), 1101–1116 (2017). https://doi.org/10.1007/s00542-016-2906-3

M. Øieroset, P.E. Sandholt, W.F. Denig, S.W.H. Cowley, Northward interplanetary magnetic field cusp aurora and high-latitude magnetopause reconnection. J. Geophys. Res. **102**, 11349–11362 (1997). https://doi.org/10.1029/97JA00559

M. Øieroset, D.L. Mitchell, T.D. Phan, R.P. Lin, M.H. Acuña, Hot diamagnetic cavities upstream of the Martian bow shock. Geophys. Res. Lett. **28**, 887–890 (2001). https://doi.org/10.1029/2000GL012289

K. Oksavik, J. Moen, H.C. Carlson, R.A. Greenwald, S.E. Milan, M. Lester, W.F. Denig, R.J. Barnes, Multi-instrument mapping of the small-scale flow dynamics related to a cusp auroral transient. Ann. Geophys. **23**, 2657–2670 (2005). https://doi.org/10.5194/angeo-23-2657-2005

N. Omidi, D.G. Sibeck, Formation of hot flow anomalies and solitary shocks. J. Geophys. Res. Space Phys. **112**, 01203 (2007). https://doi.org/10.1029/2006JA011663

N. Omidi, D. Winske, Simulation of the solar wind interaction with the outer regions of the coma. Geophys. Res. Lett. **13**, 397–400 (1986). https://doi.org/10.1029/GL013i004p00397

N. Omidi, D. Winske, A kinetic study of solar wind mass loading and cometary bow shocks. J. Geophys. Res. **92**, 13409–13426 (1987). https://doi.org/10.1029/JA092iA12p13409

N. Omidi, D.G. Sibeck, X. Blanco-Cano, Foreshock compressional boundary. J. Geophys. Res. Space Phys. **114**, 8205 (2009). https://doi.org/10.1029/2008JA013950

N. Omidi, D. Sibeck, X. Blanco-Cano, D. Rojas-Castillo, D. Turner, H. Zhang, P. Kajdič, Dynamics of the foreshock compressional boundary and its connection to foreshock cavities. J. Geophys. Res. Space Phys. **118**, 823–831 (2013). https://doi.org/10.1002/jgra.50146

M. Ong, J.G. Luhmann, C.T. Russell, R.J. Strangeway, L.H. Brace, Venus ionospheric 'clouds'—relationship to the magnetosheath field geometry. J. Geophys. Res. **96**, 11 (1991). https://doi.org/10.1029/91JA01100

T.G. Onsager, S.-W. Chang, J.D. Perez, J.B. Austin, L.X. Janoo, Low-altitude observations and modeling of quasi-steady magnetopause reconnection. J. Geophys. Res. **100**, 11831–11844 (1995). https://doi.org/10.1029/94JA02702

N. Ozak, T.E. Cravens, D.R. Schultz, Auroral ion precipitation at Jupiter: predictions for Juno. Geophys. Res. Lett. **40**, 4144–4148 (2013). https://doi.org/10.1002/grl.50812

N. Ozak, D.R. Schultz, T.E. Cravens, V. Kharchenko, Y.-W. Hui, Auroral X-ray emission at Jupiter: depth effects. J. Geophys. Res. Space Phys. **115**, 11306 (2010). https://doi.org/10.1029/2010JA015635

M. Palmroth, H. Laakso, T.I. Pulkkinen, Location of high-altitude cusp during steady solar wind conditions. J. Geophys. Res. **106**, 21109–21122 (2001). https://doi.org/10.1029/2001JA900073

E.V. Panov, J. Büchner, M. FräNz, A. Korth, S.P. Savin, H. RèMe, K.-H. FornaçOn, High-latitude Earth's magnetopause outside the cusp: cluster observations. J. Geophys. Res. Space Phys. **113**, 01220 (2008). https://doi.org/10.1029/2006JA012123

G. Parks, E. Lee, F. Mozer, N. Lin, M. Wilber, E. Lucek, Y. Dandouras, H. Reme, J. Cao, K. Meziane, C. Mazelle, M. Goldstein, P. Escoubet, Larmor radius size density holes in the solar wind upstream of the bow shock. AGU Fall Meeting Abstracts (2006)

K.I. Paularena, J.D. Richardson, M.A. Kolpak, C.R. Jackson, G.L. Siscoe, A dawn-dusk density asymmetry in Earth's magnetosheath. J. Geophys. Res. **106**, 25377–25394 (2001). https://doi.org/10.1029/2000JA000177

T. Penz, N.V. Erkaev, H.K. Biernat, H. Lammer, U.V. Amerstorfer, H. Gunell, E. Kallio, S. Barabash, S. Orsini, A. Milillo, W. Baumjohann, Ion loss on Mars caused by the Kelvin Helmholtz instability. Planet. Space Sci. **52**, 1157–1167 (2004). https://doi.org/10.1016/j.pss.2004.06.001

R. Pepino, V. Kharchenko, A. Dalgarno, R. Lallement, Spectra of the X-ray emission induced in the interaction between the solar wind and the heliospheric gas. Astrophys. J. **617**, 1347–1352 (2004). https://doi.org/10.1086/425682

S.M. Petrinec, C.T. Russell, External and internal influences on the size of the dayside terrestrial magnetosphere. Geophys. Res. Lett. **20**, 339–342 (1993). https://doi.org/10.1029/93GL00085

S.M. Petrinec, M.A. Dayeh, H.O. Funsten, S.A. Fuselier, D. Heirtzler, P. Janzen, H. Kucharek, D.J. McComas, E. Möbius, T.E. Moore, D.B. Reisenfeld, N.A. Schwadron, K.J. Trattner, P. Wurz, Neutral atom imaging of the magnetospheric cusps. J. Geophys. Res. Space Phys. **116**, 7203 (2011). https://doi.org/10.1029/2010JA016357

T.-D. Phan, G. Paschmann, W. Baumjohann, N. Sckopke, H. Luehr, The magnetosheath region adjacent to the dayside magnetopause: AMPTE/IRM observations. J. Geophys. Res. **99**, 121–141 (1994). https://doi.org/10.1029/93JA02444

T.D. Phan, L.M. Kistler, B. Klecker, G. Haerendel, G. Paschmann, B.U.Ö. Sonnerup, W. Baumjohann, M.B. Bavassano-Cattaneo, C.W. Carlson, A.M. DiLellis, K.-H. Fornacon, L.A. Frank, M. Fujimoto, E. Georgescu, S. Kokubun, E. Moebius, T. Mukai, M. Øieroset, W.R. Paterson, H. Reme, Extended magnetic reconnection at the Earth's magnetopause from detection of bi-directional jets. Nature **404**, 848–850 (2000)

T.D. Phan, G. Paschmann, J.T. Gosling, M. Oieroset, M. Fujimoto, J.F. Drake, V. Angelopoulos, The dependence of magnetic reconnection on plasma β and magnetic shear: evidence from magnetopause observations. Geophys. Res. Lett. **40**, 11–16 (2013). https://doi.org/10.1029/2012GL054528

J.L. Phillips, J.G. Luhmann, C.T. Russell, Dependence of Venus ionopause altitude and ionospheric magnetic field on solar wind dynamic pressure. Adv. Space Res. **5**, 173–176 (1985). https://doi.org/10.1016/0273-1177(85)90286-8

C.M. Pieters, J.N. Goswami, R.N. Clark, M. Annadurai, J. Boardman, B. Buratti, J.-P. Combe, M.D. Dyar, R. Green, J.W. Head, C. Hibbitts, M. Hicks, P. Isaacson, R. Klima, G. Kramer, S. Kumar, E. Livo, S. Lundeen, E. Malaret, T. McCord, J. Mustard, J. Nettles, N. Petro, C. Runyon, M. Staid, J. Sunshine, L.A. Taylor, S. Tompkins, P. Varanasi, Character and spatial distribution of OH/H^2O on the surface of the Moon seen by M^3 on Chandrayaan-1. Science **326**, 568 (2009). https://doi.org/10.1126/science.1178658

M. Pinnock, A.S. Rodger, J.R. Dudeney, K.B. Baker, P.T. Neweli, R.A. Greenwald, M.E. Greenspan, Observations of an enhanced convection channel in the cusp ionosphere. J. Geophys. Res. **98**, 3767–3776 (1993). https://doi.org/10.1029/92JA01382

F. Pitout, C.P. Escoubet, B. Klecker, H. Rème, Cluster survey of the mid-altitude cusp: 1. Size, location, and dynamics. Ann. Geophys. **24**, 3011–3026 (2006). https://doi.org/10.5194/angeo-24-3011-2006

V. Pizzo, A three-dimensional model of corotating streams in the solar wind. I—theoretical foundations. J. Geophys. Res. **83**, 5563–5572 (1978). https://doi.org/10.1029/JA083iA12p05563

S.A. Pope, M.A. Balikhin, T.L. Zhang, A.O. Fedorov, M. Gedalin, S. Barabash, Giant vortices lead to ion escape from Venus and re-distribution of plasma in the ionosphere. Geophys. Res. Lett. **36**, 07202 (2009). https://doi.org/10.1029/2008GL036977

A.E. Potter, R.M. Killen, T.H. Morgan, Variation of lunar sodium during passage of the Moon through the Earth's magnetotail. J. Geophys. Res. **105**, 15073–15084 (2000). https://doi.org/10.1029/1999JE001213

M.I. Pudovkin, S.A. Zaitseva, B.P. Besser, Magnetopause magnetic barrier parameters in dependence on the solar wind magnetic field orientation. Ann. Geophys. **13**, 828–835 (1995). https://doi.org/10.1007/s00585-995-0828-y

M.I. Pudovkin, B.P. Besser, S.A. Zaitseva, V.V. Lebedeva, C.-V. Meister, Magnetic barrier in case of a southward interplanetary magnetic field. J. Atmos. Sol.-Terr. Phys. **63**, 1075–1083 (2001). https://doi.org/10.1016/S1364-6826(01)00023-2

M.I. Pudovkin, S.A. Zaitseva, V.V. Lebedeva, A.A. Samsonov, B.P. Besser, C.-V. Meister, W. Baumjohann, MHD-modelling of the magnetosheath. Planet. Space Sci. **50**, 473–488 (2002). https://doi.org/10.1016/S0032-0633(02)00027-2

E. Quémerais, R. Lallement, S. Ferron, D. Koutroumpa, J.-L. Bertaux, E. Kyrölä, W. Schmidt, Interplanetary hydrogen absolute ionization rates: retrieving the solar wind mass flux latitude and cycle dependence with SWAN/SOHO maps. J. Geophys. Res. Space Phys. **111**, 09114 (2006). https://doi.org/10.1029/2006JA011711

W. Raab, G. Branduardi-Raymont, C. Wang, L. Dai, E. Donovan, G. Enno, P. Escoubet, A. Holland, L. Jing, D. Kataria, L. Li, A. Read, D. Rebuffat, J. Romstedt, C. Runciman, S. Sembay, E. Spanswick, J. Sykes, J. Thornhill, A. Wielders, A. Zhang, J. Zheng, SMILE: a joint ESA/CAS mission to investigate the interaction between the solar wind and Earth's magnetosphere, in *Space Telescopes and Instrumentation 2016: Ultraviolet to Gamma Ray*. Proc. SPIE., vol. 9905 (2016), p. 990502. https://doi.org/10.1117/12.2231984

P.H. Reiff, J.L. Burch, T.W. Hill, Solar wind plasma injection at the dayside magnetospheric cusp. J. Geophys. Res. **82**, 479–491 (1977). https://doi.org/10.1029/JA082i004p00479

A.J. Ridley, Alfvén wings at Earth's magnetosphere under strong interplanetary magnetic fields. Ann. Geophys. **25**, 533–542 (2007). https://doi.org/10.5194/angeo-25-533-2007

I.P. Robertson, T.E. Cravens, Spatial maps of heliospheric and geocoronal X-ray intensities due to the charge exchange of the solar wind with neutrals. J. Geophys. Res. Space Phys. **108**, 8031 (2003a). https://doi.org/10.1029/2003JA009873

I.P. Robertson, T.E. Cravens, X-ray emission from the terrestrial magnetosheath. Geophys. Res. Lett. **30**, 1439 (2003b). https://doi.org/10.1029/2002GL016740

I.P. Robertson, T.E. Cravens, M.V. Medvedev, M.R. Collier, G.P. Zank, V. Florinski, X-ray emissions from charge exchange in the heliosphere, in *Solar Wind 11/SOHO 16, Connecting Sun and Heliosphere*, ed. by B. Fleck, T.H. Zurbuchen, H. Lacoste. ESA Special Publication, vol. 592 (2005), pp. 41–45

I.P. Robertson, M.R. Collier, T.E. Cravens, M.-C. Fok, X-ray emission from the terrestrial magnetosheath including the cusps. J. Geophys. Res. Space Phys. **111**, 12105 (2006). https://doi.org/10.1029/2006JA011672

I.P. Robertson, S. Sembay, T.J. Stubbs, K.D. Kuntz, M.R. Collier, T.E. Cravens, S.L. Snowden, H.K. Hills, F.S. Porter, P. Travnicek, J.A. Carter, A.M. Read, Solar wind charge exchange observed through the lunar exosphere. Geophys. Res. Lett. **36**, 21102 (2009a). https://doi.org/10.1029/2009GL040834

I.P. Robertson, K.D. Kuntz, M.R. Collier, T.E. Cravens, S.L. Snowden, The heliospheric contribution to the soft X-ray background emission, in *American Institute of Physics Conference Series*, ed. by R.K. Smith, S.L. Snowden, K.D. Kuntz. American Institute of Physics Conference Series, vol. 1156 (2009b), pp. 52–61. https://doi.org/10.1063/1.3211834

I.P. Robertson, T.E. Cravens, D.G. Sibeck, M.R. Collier, K.D. Kuntz, Solar wind charge exchange during geomagnetic storms. Astron. Nachr. **333**, 309 (2012). https://doi.org/10.1002/asna.201211671

I.P. Robertson, T.E. Cravens, M.R. Collier, D.G. Sibeck, K.D. Kuntz, S.L. Snowden, Solar wind charge exchange and Earth's magnetosheath, in *American Institute of Physics Conference Series*, ed. by G.P. Zank, J. Borovsky, R. Bruno, J. Cirtain, S. Cranmer, H. Elliott, J. Giacalone, W. Gonzalez, G. Li, E. Marsch, E. Moebius, N. Pogorelov, J. Spann, O. Verkhoglyadova. American Institute of Physics Conference Series, vol. 1539 (2013), pp. 430–433. https://doi.org/10.1063/1.4811077

E.C. Roelof, D.G. Sibeck, Magnetopause shape as a bivariate function of interplanetary magnetic field B_z and solar wind dynamic pressure. J. Geophys. Res. **98**, 21421 (1993). https://doi.org/10.1029/93JA02362

H. Rosenbauer, H. Gruenwaldt, M.D. Montgomery, G. Paschmann, N. Sckopke, Heos 2 plasma observations in the distant polar magnetosphere—the plasma mantle. J. Geophys. Res. **80**, 2723–2737 (1975). https://doi.org/10.1029/JA080i019p02723

C.L. Rufenach, H.H. Sauer, R.F. Martin Jr., A study of geosynchronous magnetopause crossings. J. Geophys. Res. **94**, 15125–15134 (1989). https://doi.org/10.1029/JA094iA11p15125

C.T. Russell, R.C. Elphic, Initial ISEE magnetometer results—magnetopause observations. Space Sci. Rev. **22**, 681–715 (1978). https://doi.org/10.1007/BF00212619

C.T. Russell, R.C. Elphic, Observation of magnetic flux ropes in the Venus ionosphere. Nature **279**, 616–618 (1979). https://doi.org/10.1038/279616a0

C.T. Russell, T. Mulligan, On the magnetosheath thicknesses of interplanetary coronal mass ejections. Planet. Space Sci. **50**, 527–534 (2002). https://doi.org/10.1016/S0032-0633(02)00031-4

C.T. Russell, J.G. Luhmann, R.J. Strangeway, The solar wind interaction with Venus through the eyes of the Pioneer Venus Orbiter. Planet. Space Sci. **54**, 1482–1495 (2006). https://doi.org/10.1016/j.pss.2006.04.025

C.T. Russell, M.M. Hoppe, W.A. Livesey, J.T. Gosling, S.J. Bame, ISEE-1 and -2 observations of laminar bow shocks—velocity and thickness. Geophys. Res. Lett. **9**, 1171–1174 (1982a). https://doi.org/10.1029/GL009i010p01171

C.T. Russell, J.G. Luhmann, R.C. Elphic, F.L. Scarf, L.H. Brace, Magnetic field and plasma wave observations in a plasma cloud at Venus. Geophys. Res. Lett. **9**, 45–48 (1982b). https://doi.org/10.1029/GL009i001p00045

C.T. Russell, J.T. Gosling, R.D. Zwickl, E.J. Smith, Multiple spacecraft observations of interplanetary shocks ISEE three-dimensional plasma measurements. J. Geophys. Res. **88**, 9941–9947 (1983). https://doi.org/10.1029/JA088iA12p09941

C.T. Russell, R.N. Singh, J.G. Luhmann, R.C. Elphic, L.H. Brace, Waves on the subsolar ionopause of Venus. Adv. Space Res. **7**, 115–118 (1987). https://doi.org/10.1016/0273-1177(87)90209-2

C.T. Russell, E. Chou, J.G. Luhmann, P. Gazis, L.H. Brace, W.R. Hoegy, Solar and interplanetary control of the location of the Venus bow shock. J. Geophys. Res. **93**, 5461–5469 (1988). https://doi.org/10.1029/JA093iA06p05461

C.T. Russell, S.M. Petrinec, T.L. Zhang, P. Song, H. Kawano, The effect of foreshock on the motion of the dayside magnetopause. Geophys. Res. Lett. **24**, 1439–1441 (1997). https://doi.org/10.1029/97GL01408

A.A. Samsonov, D. Hubert, Steady state slow shock inside the Earth's magnetosheath: to be or not to be? 2. Numerical three-dimensional MHD modeling. J. Geophys. Res. Space Phys. **109**, 01218 (2004). https://doi.org/10.1029/2003JA010006

A.A. Samsonov, Z. Němeček, J. Šafránková, K. Jelínek, Why does the subsolar magnetopause move sunward for radial interplanetary magnetic field? J. Geophys. Res. Space Phys. **117**, 5221 (2012). https://doi.org/10.1029/2011JA017429

A.A. Samsonov, Z. Němeček, J. Šafránková, K. Jelínek, Why does the total pressure on the subsolar magnetopause differ from the solar wind dynamic pressure? Cosm. Res. **51**, 37–45 (2013). https://doi.org/10.1134/S0010952513010073

A.A. Samsonov, V.A. Sergeev, M.M. Kuznetsova, D.G. Sibeck, Asymmetric magnetospheric compressions and expansions in response to impact of inclined interplanetary shock. Geophys. Res. Lett. **42**, 4716–4722 (2015). https://doi.org/10.1002/2015GL064294

A.A. Samsonov, E. Gordeev, N.A. Tsyganenko, J. Šafránková, Z. Němeček, J. Šimunek, D.G. Sibeck, G. Tóth, V.G. Merkin, J. Raeder, Do we know the actual magnetopause position for typical solar wind conditions? J. Geophys. Res. Space Phys. **121**, 6493–6508 (2016). https://doi.org/10.1002/2016JA022471

A.A. Samsonov, D.G. Sibeck, J. Å afránková, Z. Němeček, J.-H. Shue, A method to predict magnetopause expansion in radial IMF events by MHD simulations. J. Geophys. Res. Space Phys. **122**, 3110–3126 (2017). https://doi.org/10.1002/2016JA023301

W.T. Sanders, W.L. Kraushaar, J.A. Nousek, P.M. Fried, Soft diffuse X-rays in the southern galactic hemisphere. Astrophys. J. Lett. **217**, 87–91 (1977). https://doi.org/10.1086/182545

P.E. Sandholt, C.J. Farrugia, Does the aurora provide evidence for the occurrence of antiparallel magnetopause reconnection? J. Geophys. Res. Space Phys. **108**, 1466 (2003). https://doi.org/10.1029/2003JA010066

P.E. Sandholt, C.J. Farrugia, M. Øieroset, P. Stauning, W.F. Denig, Auroral activity associated with unsteady magnetospheric erosion: observations on December 18, 1990. J. Geophys. Res. **103**, 2309–2318 (1998). https://doi.org/10.1029/97JA01317

M. Sarantos, R.M. Killen, D.A. Glenar, M. Benna, T.J. Stubbs, Metallic species, oxygen and silicon in the lunar exosphere: upper limits and prospects for LADEE measurements. J. Geophys. Res. Space Phys. **117**, 03103 (2012). https://doi.org/10.1029/2011JA017044

M.A. Schield, Pressure balance between solar wind and magnetosphere. J. Geophys. Res. **74**, 1275 (1969). https://doi.org/10.1029/JA074i005p01275

J.H.M.M. Schmitt, H. Fink, F.R. Harnden Jr., X-ray observations of solar flares with the Einstein Observatory. Astrophys. J. **322**, 1023–1034 (1987). https://doi.org/10.1086/165797

J.H.M.M. Schmitt, S.L. Snowden, B. Aschenbach, G. Hasinger, E. Pfeffermann, P. Predehl, J. Trumper, A soft X-ray image of the Moon. Nature **349**, 583–587 (1991). https://doi.org/10.1038/349583a0

P.H. Schultz, B. Hermalyn, A. Colaprete, K. Ennico, M. Shirley, W.S. Marshall, The LCROSS cratering experiment. Science **330**, 468 (2010). https://doi.org/10.1126/science.1187454

N.A. Schwadron, T.E. Cravens, Implications of solar wind composition for cometary X-rays. Astrophys. J. **544**, 558–566 (2000). https://doi.org/10.1086/317176

N.A. Schwadron, L.A. Fisk, T.H. Zurbuchen, Elemental fractionation in the slow solar wind. Astrophys. J. **521**, 859–867 (1999). https://doi.org/10.1086/307575

L.S. Shepherd, P.A. Cassak, Guide field dependence of 3-D X-line spreading during collisionless magnetic reconnection. J. Geophys. Res. Space Phys. **117**, 10101 (2012). https://doi.org/10.1029/2012JA017867

H. Shimazu, Effects of charge exchange and photoionization on the interaction between the solar wind and unmagnetized planets. J. Geophys. Res. **106**, 18751–18762 (2001a). https://doi.org/10.1029/2001JA900010

H. Shimazu, Three-dimensional hybrid simulation of solar wind interaction with unmagnetized planets. J. Geophys. Res. **106**, 8333–8342 (2001b). https://doi.org/10.1029/2000JA900069

J.-H. Shue, P. Song, C.T. Russell, J.T. Steinberg, J.K. Chao, G. Zastenker, O.L. Vaisberg, S. Kokubun, H.J. Singer, T.R. Detman, H. Kawano, Magnetopause location under extreme solar wind conditions. J. Geophys. Res. **103**, 17691–17700 (1998). https://doi.org/10.1029/98JA01103

J.-H. Shue, P. Song, C.T. Russell, J.K. Chao, Y.-H. Yang, Toward predicting the position of the magnetopause within geosynchronous orbit. J. Geophys. Res. **105**, 2641–2656 (2000). https://doi.org/10.1029/1999JA900467

J.-H. Shue, P. Song, C.T. Russell, M.F. Thomsen, S.M. Petrinec, Dependence of magnetopause erosion on southward interplanetary magnetic field. J. Geophys. Res. **106**, 18777–18788 (2001). https://doi.org/10.1029/2001JA900039

J. Shue, Y. Hsieh, W.Y. Tam, K. Wang, H. Fu, J. Bortnik, Invariant distribution of repetition period for Whistler-mode rising-tone chorus. AGU Fall Meeting Abstracts (2013)

D.G. Sibeck, R.E. Lopez, E.C. Roelof, Solar wind control of the magnetopause shape, location, and motion. J. Geophys. Res. **96**, 5489–5495 (1991). https://doi.org/10.1029/90JA02464

D.G. Sibeck, R.W. McEntire, K.S. Ross, Charge states of substorm particle injections. Geophys. Res. Lett. **15**, 1283–1286 (1988). https://doi.org/10.1029/GL015i011p01283

D.G. Sibeck, R.B. Decker, D.G. Mitchell, A.J. Lazarus, R.P. Lepping, A. Szabo, Solar wind preconditioning in the flank foreshock: IMP 8 observations. J. Geophys. Res. **106**, 21675–21688 (2001). https://doi.org/10.1029/2000JA000417

D.G. Sibeck, G.I. Korotova, V. Petrov, V. Styazhkin, T.J. Rosenberg, Flux transfer events on the high-latitude magnetopause: interball-1 observations. Ann. Geophys. **23**, 3549–3559 (2005). https://doi.org/10.5194/angeo-23-3549-2005

D.G. Sibeck, N. Omidi, I. Dandouras, E. Lucek, On the edge of the foreshock: model-data comparisons. Ann. Geophys. **26**, 1539–1544 (2008). https://doi.org/10.5194/angeo-26-1539-2008

C. Simon Wedlund, M. Alho, G. Gronoff, E. Kallio, H. Gunell, H. Nilsson, J. Lindkvist, E. Behar, G. Stenberg Wieser, W.J. Miloch, Hybrid modelling of cometary plasma environments. I. Impact of photoionisation, charge exchange, and electron ionisation on bow shock and cometopause at 67P/Churyumov-Gerasimenko. Astron. Astrophys. **604**, 73 (2017). https://doi.org/10.1051/0004-6361/201730514

G. Siscoe, J. Raeder, A.J. Ridley, Transpolar potential saturation models compared. J. Geophys. Res. Space Phys. **109**, 9203 (2004). https://doi.org/10.1029/2003JA010318

G.L. Siscoe, N.U. Crooker, G.M. Erickson, B.U.Ö. Sonnerup, N.C. Maynard, J.A. Schoendorf, K.D. Siebert, D.R. Weimer, W.W. White, G.R. Wilson, MHD properties of magnetosheath flow. Planet. Space Sci. **50**, 461–471 (2002). https://doi.org/10.1016/S0032-0633(02)00026-0

J.A. Slavin, R.C. Elphic, C.T. Russell, F.L. Scarf, J.H. Wolfe, J.D. Mihalov, D.S. Intriligator, L.H. Brace, H.A. Taylor, R.E. Daniell, The solar wind interaction with Venus—Pioneer Venus observations of bow shock location and structure. J. Geophys. Res. **85**, 7625–7641 (1980). https://doi.org/10.1029/JA085iA13p07625

J.A. Slavin, A. Szabo, M. Peredo, R.P. Lepping, R.J. Fitzenreiter, K.W. Ogilvie, C.J. Owen, J.T. Steinberg, Near-simultaneous bow shock crossings by WIND and IMP 8 on December 1, 1994. Geophys. Res. Lett. **23**, 1207–1210 (1996). https://doi.org/10.1029/96GL01351

K.Y. Slivka, V.S. Semenov, N.V. Erkaev, N.P. Dmitrieva, I.V. Kubyshkin, H. Lammer, Peculiarities of magnetic barrier formation for southward and northward directions of the IMF. J. Geophys. Res. Space Phys. **120**, 9471–9483 (2015). https://doi.org/10.1002/2015JA021250

M.F. Smith, M. Lockwood, The pulsating cusp. Geophys. Res. Lett. **17**, 1069–1072 (1990). https://doi.org/10.1029/GL017i008p01069

R.K. Smith, R.J. Edgar, P.P. Plucinsky, B.J. Wargelin, P.E. Freeman, B.A. Biller, Chandra observations of MBM 12 and models of the Local Bubble. Astrophys. J. **623**, 225–234 (2005). https://doi.org/10.1086/428568

R.K. Smith, A.R. Foster, R.J. Edgar, N.S. Brickhouse, Resolving the origin of the diffuse soft X-ray background. Astrophys. J. **787**, 77 (2014). https://doi.org/10.1088/0004-637X/787/1/77

S.L. Snowden, The hot part of the interstellar medium, in *The Century of Space Science*, ed. by J.A. Bleeker, J. Geiss, M.C.E. Huber (Kluwer Academic Publishers, Dordrecht, 2002), p. 581

S.L. Snowden, M.J. Freyberg, The scattered solar X-ray background of the ROSAT PSPC. Astrophys. J. **404**, 403–411 (1993). https://doi.org/10.1086/172289

S.L. Snowden, R. Petre, An X-ray image of the large Magellanic cloud. Astrophys. J. Lett. **436**, 123–126 (1994). https://doi.org/10.1086/187648

S.L. Snowden, M.R. Collier, K.D. Kuntz, XMM-Newton observation of solar wind charge exchange emission. Astrophys. J. **610**, 1182–1190 (2004). https://doi.org/10.1086/421841

S.L. Snowden, D.P. Cox, D. McCammon, W.T. Sanders, A model for the distribution of material generating the soft X-ray background. Astrophys. J. **354**, 211–219 (1990). https://doi.org/10.1086/168680

S.L. Snowden, D. McCammon, D.N. Burrows, J.A. Mendenhall, Analysis procedures for ROSAT XRT/PSPC observations of extended objects and the diffuse background. Astrophys. J. **424**, 714–728 (1994). https://doi.org/10.1086/173925

S.L. Snowden, M.J. Freyberg, P.P. Plucinsky, J.H.M.M. Schmitt, J. Truemper, W. Voges, R.J. Edgar, D. McCammon, W.T. Sanders, First maps of the soft X-ray diffuse background from the ROSAT XRT/PSPC all-sky survey. Astrophys. J. **454**, 643 (1995). https://doi.org/10.1086/176517

S.L. Snowden, R. Egger, M.J. Freyberg, D. McCammon, P.P. Plucinsky, W.T. Sanders, J.H.M.M. Schmitt, J. Truemper, W. Voges, ROSAT survey diffuse X-ray background maps. II. Astrophys. J. **485**, 125 (1997). https://doi.org/10.1086/304399

S.L. Snowden, M.R. Collier, T. Cravens, K.D. Kuntz, S.T. Lepri, I. Robertson, L. Tomas, Observation of solar wind charge exchange emission from exospheric material in and outside Earth's magnetosheath. Astrophys. J. **691**, 372–381 (2009a). https://doi.org/10.1088/0004-637X/691/1/372

S.L. Snowden, K.D. Kuntz, M.R. Collier, M.R. Collier, SWCX emission from the helium focusing cone-preliminary results, in *American Institute of Physics Conference Series*, ed. by R.K. Smith, S.L. Snowden, K.D. Kuntz. American Institute of Physics Conference Series, vol. 1156 (2009b), pp. 90–94. https://doi.org/10.1063/1.3211838

S.L. Snowden, M. Chiao, M.R. Collier, F.S. Porter, N.E. Thomas, T. Cravens, I.P. Robertson, M. Galeazzi, Y. Uprety, E. Ursino, D. Koutroumpa, K.D. Kuntz, R. Lallement, L. Puspitarini, S.T. Lepri, D. McCammon, K. Morgan, B.M. Walsh, Pressure equilibrium between the local interstellar clouds and the local hot bubble. Astrophys. J. Lett. **791**, 14 (2014). https://doi.org/10.1088/2041-8205/791/1/L14

S.L. Snowden, D. Koutroumpa, K.D. Kuntz, R. Lallement, L. Puspitarini, The north galactic pole rift and the local hot bubble. Astrophys. J. **806**, 120 (2015). https://doi.org/10.1088/0004-637X/806/1/120

C.V. Solodyna, J.W. Belcher, J.W. Sari, Plasma field characteristics of directional discontinuities in the inter-planetary medium. J. Geophys. Res. **82**, 10–14 (1977). https://doi.org/10.1029/JA082i001p00010

P. Song, C.T. Russell, Model of the formation of the low-latitude boundary layer for strongly northward inter-planetary magnetic field. J. Geophys. Res. **97**, 1411–1420 (1992). https://doi.org/10.1029/91JA02377

P. Song, C.T. Russell, J.T. Gosling, M. Thomsen, R.C. Elphic, Observations of the density profile in the magnetosheath near the stagnation streamline. Geophys. Res. Lett. **17**, 2035–2038 (1990). https://doi.org/10.1029/GL017i011p02035

P. Song, C.T. Russell, J.T. Gosling, M.F. Thomsen, R.C. Elphic, Comment on "Steady state slow shock inside the Earth's magnetosheath: to be or not to be? 1. The original observation revisited" by D. Hubert and A. Samsonov. J. Geophys. Res. Space Phys. **110**, 11210 (2005). https://doi.org/10.1029/2005JA011161

B.U.O. Sonnerup, G. Paschmann, I. Papamastorakis, N. Sckopke, G. Haerendel, S.J. Bame, J.R. Asbridge, J.T. Gosling, C.T. Russell, Evidence for magnetic field reconnection at the Earth's magnetopause. J. Geophys. Res. **86**, 10049–10067 (1981). https://doi.org/10.1029/JA086iA12p10049

D.J. Southwood, The hydromagnetic stability of the magnetospheric boundary. Planet. Space Sci. **16**, 587–605 (1968). https://doi.org/10.1016/0032-0633(68)90100-1

D.J. Southwood, M.G. Kivelson, On the form of the flow in the magnetosheath. J. Geophys. Res. **97**, 2873–2879 (1992). https://doi.org/10.1029/91JA02446

D.J. Southwood, M.G. Kivelson, Magnetosheath flow near the subsolar magnetopause: Zwan-Wolf and Southwood-Kivelson theories reconciled. Geophys. Res. Lett. **22**, 3275–3278 (1995). https://doi.org/10.1029/95GL03131

W.N. Spjeldvik, T.A. Fritz, Theory for charge states of energetic oxygen ions in the Earth's radiation belts. J. Geophys. Res. **83**, 1583–1594 (1978). https://doi.org/10.1029/JA083iA04p01583

J.R. Spreiter, A.L. Summers, A.Y. Alksne, Hydromagnetic flow around the magnetosphere. Planet. Space Sci. **14**, 223 (1966). https://doi.org/10.1016/0032-0633(66)90124-3

S.A. Stern, The lunar atmosphere: history, status, current problems, and context. Rev. Geophys. **37**, 453–492 (1999). https://doi.org/10.1029/1999RG900005

W. Stüdemann, B. Wilken, D.N. Baker, P.R. Higbie, R.D. Belian, Multispacecraft observations at the compressed magnetopause following the 13 July 1982 interplanetary shock. Planet. Space Sci. **34**, 825–833 (1986). https://doi.org/10.1016/0032-0633(86)90079-6

T.R. Sun, C. Wang, F. Wei, S. Sembay, X-ray imaging of Kelvin-Helmholtz waves at the magnetopause. J. Geophys. Res. Space Phys. **120**, 266–275 (2015). https://doi.org/10.1002/2014JA020497

J.M. Sunshine, T.L. Farnham, L.M. Feaga, O. Groussin, F. Merlin, R.E. Milliken, M.F. A'Hearn, Temporal and spatial variability of lunar hydration as observed by the Deep Impact Spacecraft. Science **326**, 565 (2009). https://doi.org/10.1126/science.1179788

A.V. Suvorova, J.-H. Shue, A.V. Dmitriev, D.G. Sibeck, J.P. McFadden, H. Hasegawa, K. Ackerson, K. Jelínek, J. Šafránková, Z. Němeček, Magnetopause expansions for quasi-radial interplanetary magnetic field: THEMIS and geotail observations. J. Geophys. Res. Space Phys. **115**, 10216 (2010). https://doi.org/10.1029/2010JA015404

S. Taguchi, M.R. Collier, T.E. Moore, M.-C. Fok, H.J. Singer, Response of neutral atom emissions in the low-latitude and high-latitude magnetosheath direction to the magnetopause motion under extreme solar wind conditions. J. Geophys. Res. Space Phys. **109**, 04208 (2004). https://doi.org/10.1029/2003JA010147

T. Takeuchi, C.T. Russell, T. Araki, Effect of the orientation of interplanetary shock on the geomagnetic sudden commencement. J. Geophys. Res. Space Phys. **107**, 1423 (2002). https://doi.org/10.1029/2002JA009597

Y. Tanaka, J.A.M. Bleeker, The diffuse soft X-ray sky—astrophysics related to cosmic soft X-rays in the energy range 0.1–2.0 keV. Space Sci. Rev. **20**, 815–888 (1977). https://doi.org/10.1007/BF02431836

M. Tatrallyay, C.T. Russell, J.D. Mihalov, A. Barnes, Factors controlling the location of the Venus bow shock. J. Geophys. Res. **88**, 5613–5621 (1983). https://doi.org/10.1029/JA088iA07p05613

M. Tatrallyay, C.T. Russell, J.G. Luhmann, A. Barnes, J.D. Mihalov, On the proper Mach number and ratio of specific heats for modeling the Venus bow shock. J. Geophys. Res. **89**, 7381–7392 (1984). https://doi.org/10.1029/JA089iA09p07381

M.G.G.T. Taylor, H. Hasegawa, B. Lavraud, T. Phan, C.P. Escoubet, M.W. Dunlop, Y.V. Bogdanova, A.L. Borg, M. Volwerk, J. Berchem, O.D. Constantinescu, J.P. Eastwood, A. Masson, H. Laakso, J. Soucek, A.N. Fazakerley, H.U. Frey, E.V. Panov, C. Shen, J.K. Shi, D.G. Sibeck, Z.Y. Pu, J. Wang, J.A. Wild, Spatial distribution of rolled up Kelvin-Helmholtz vortices at Earth's dayside and flank magnetopause. Ann. Geophys. **30**, 1025–1035 (2012)

V. Tenishev, M. Rubin, O.J. Tucker, M.R. Combi, M. Sarantos, Kinetic modeling of sodium in the lunar exosphere. Icarus **226**, 1538–1549 (2013). https://doi.org/10.1016/j.icarus.2013.08.021

120 ✐ Springer

N. Terada, S. Machida, H. Shinagawa, Global hybrid simulation of the Kelvin-Helmholtz instability at the Venus ionopause. J. Geophys. Res. Space Phys. **107**, 1471 (2002). https://doi.org/10.1029/2001JA009224

N. Terada, H. Shinagawa, T. Tanaka, K. Murawski, K. Terada, A three-dimensional, multispecies, comprehensive MHD model of the solar wind interaction with the planet Venus. J. Geophys. Res. Space Phys. **114**, 09208 (2009). https://doi.org/10.1029/2008JA013937

N.E. Thomas, J.A. Carter, M.P. Chiao, D.J. Chornay, Y.M. Collado-Vega, M.R. Collier, T.E. Cravens, M. Galeazzi, D. Koutroumpa, J. Kujawski, K.D. Kuntz, M.M. Kuznetsova, S.T. Lepri, D. McCammon, K. Morgan, F.S. Porter, K. Prasai, A.M. Read, I.P. Robertson, S.F. Sembay, D.G. Sibeck, S.L. Snowden, Y. Uprety, B.M. Walsh, The DXL and STORM sounding rocket mission, in *UV, X-Ray, and Gamma-Ray Space Instrumentation for Astronomy XVIII*. Proc. SPIE., vol. 8859 (2013), p. 88590. https://doi.org/10.1117/12.2024438

V.A. Thomas, S.H. Brecht, Evolution of diamagnetic cavities in the solar wind. J. Geophys. Res. **93**, 11341–11353 (1988). https://doi.org/10.1029/JA093iA10p11341

M.F. Thomsen, J.T. Gosling, S.A. Fuselier, S.J. Bame, C.T. Russell, Hot, diamagnetic cavities upstream from the Earth's bow shock. J. Geophys. Res. **91**, 2961–2973 (1986). https://doi.org/10.1029/JA091iA03p02961

D.V. Titov, H. Svedhem, D. Koschny, R. Hoofs, S. Barabash, J.-L. Bertaux, P. Drossart, V. Formisano, B. Häusler, O. Korablev, W.J. Markiewicz, D. Nevejans, M. Pätzold, G. Piccioni, T.L. Zhang, D. Merritt, O. Witasse, J. Zender, A. Accomazzo, M. Sweeney, D. Trillard, M. Janvier, A. Clochet, Venus express science planning. Planet. Space Sci. **54**, 1279–1297 (2006). https://doi.org/10.1016/j.pss.2006.04.017

O. Tkachenko, J. Šafránková, Z. Němeček, D.G. Sibeck, Dayside magnetopause transients correlated with changes of the magnetosheath magnetic field orientation. Ann. Geophys. **29**, 687–699 (2011). https://doi.org/10.5194/angeo-29-687-2011

C.R. Tooley, M.B. Houghton, R.S. Saylor, C. Peddie, D.F. Everett, C.L. Baker, K.N. Safdie, Lunar reconnaissance orbiter mission and spacecraft design. Space Sci. Rev. **150**, 23–62 (2010). https://doi.org/10.1007/s11214-009-9624-4

M.R. Torr, D.G. Torr, M. Zukic, R.B. Johnson, J. Ajello, P. Banks, K. Clark, K. Cole, C. Keffer, G. Parks, B. Tsurutani, J. Spann, A far ultraviolet imager for the international solar-terrestrial physics mission. Space Sci. Rev. **71**, 329–383 (1995). https://doi.org/10.1007/BF00751335

G. Tóth, B. van der Holst, I.V. Sokolov, D.L. De Zeeuw, T.I. Gombosi, F. Fang, W.B. Manchester, X. Meng, D. Najib, K.G. Powell, Q.F. Stout, A. Glocer, Y.-J. Ma, M. Opher, Adaptive numerical algorithms in space weather modeling. J. Comput. Phys. **231**, 870–903 (2012). https://doi.org/10.1016/j.jcp.2011.02.006

K.J. Trattner, S.A. Fuselier, W.K. Peterson, J.-A. Sauvaud, H. Stenuit, N. Dubouloz, R.A. Kovrazhkin, On spatial and temporal structures in the cusp. J. Geophys. Res. **104**, 28411–28422 (1999). https://doi.org/10.1029/1999JA900419

K.J. Trattner, S.A. Fuselier, S.M. Petrinec, T.K. Yeoman, C.P. Escoubet, H. Reme, The reconnection site of temporal cusp structures. J. Geophys. Res. Space Phys. **113**, 7 (2008). https://doi.org/10.1029/2007JA012776

J.G. Trotignon, E. Dubinin, R. Grard, S. Barabash, R. Lundin, Martian planetopause as seen by the plasma wave system onboard Phobos 2. J. Geophys. Res. **101**, 24965–24978 (1996). https://doi.org/10.1029/96JA01898

J. Trümper, ROSAT—a new look at the X-ray sky (the 1991 Grubb-Parsons lecture). Q. J. R. Astron. Soc. **33**, 165–174 (1992)

N.A. Tsyganenko, C.T. Russell, Magnetic signatures of the distant polar cusps: observations by polar and quantitative modeling. J. Geophys. Res. **104**, 24939–24956 (1999). https://doi.org/10.1029/1999JA900279

N.A. Tsyganenko, D.G. Sibeck, Concerning flux erosion from the dayside magnetosphere. J. Geophys. Res. **99**, 13425 (1994). https://doi.org/10.1029/94JA00719

D.L. Turner, N. Omidi, D.G. Sibeck, V. Angelopoulos, First observations of foreshock bubbles upstream of Earth's bow shock: characteristics and comparisons to HFAs. J. Geophys. Res. Space Phys. **118**, 1552–1570 (2013). https://doi.org/10.1002/jgra.50198

Y. Uprety, M. Chiao, M.R. Collier, T. Cravens, M. Galeazzi, D. Koutroumpa, K.D. Kuntz, R. Lallement, S.T. Lepri, W. Liu, D. McCammon, K. Morgan, F.S. Porter, K. Prasai, S.L. Snowden, N.E. Thomas, E. Ursino, B.M. Walsh, Solar wind charge exchange contribution to the ROSAT all sky survey maps. ArXiv e-prints (2016)

P. van Loevezijn, R. Schlatmann, J. Verhoeven, B.A. van Tiggelen, E.M. Gullikson, Numerical and experimental study of disordered multilayers for broadband x-ray reflection. Appl. Opt. **35**, 3614–3619 (1996). https://doi.org/10.1364/AO.35.003614

M. Verigin, G. Kotova, A. Szabo, J. Slavin, T. Gombosi, K. Kabin, F. Shugaev, A. Kalinchenko, Wind observations of the terrestrial bow shock: 3-D shape and motion. Earth Planets Space **53**, 1001–1009 (2001)

D. Vignes, C. Mazelle, H. Rme, M.H. Acuña, J.E.P. Connerney, R.P. Lin, D.L. Mitchell, P. Cloutier, D.H. Crider, N.F. Ness, The solar wind interaction with Mars: locations and shapes of the bow shock and the magnetic pile-up boundary from the observations of the MAG/ER experiment onboard Mars global surveyor. Geophys. Res. Lett. **27**, 49–52 (2000). https://doi.org/10.1029/1999GL010703

D. Vignes, M.H. Acuna, D.H. Crider, J. Connerney, H. Reme, C. Mazelle, Factors controlling the location of the Bow Shock at Mars. AGU Spring Meeting Abstracts (2002)

D. Vignes, M.H. Acuña, J.E.P. Connerney, D.H. Crider, H. Rème, C. Mazelle, Magnetic flux ropes in the Martian atmosphere: global characteristics. Space Sci. Rev. **111**, 223–231 (2004). https://doi.org/10.1023/B:SPAC.0000032716.21619.f2

W. Voges, B. Aschenbach, T. Boller, H. Bräuninger, U. Briel, W. Burkert, K. Dennerl, J. Englhauser, R. Gruber, F. Haberl, G. Hartner, G. Hasinger, M. Kürster, E. Pfeffermann, W. Pietsch, P. Predehl, C. Rosso, J.H.M.M. Schmitt, J. Trümper, H.U. Zimmermann, The ROSAT all-sky survey bright source catalogue. Astron. Astrophys. **349**, 389–405 (1999)

M.F. Vogt, M.G. Kivelson, K.K. Khurana, R.J. Walker, B. Bonfond, D. Grodent, A. Radioti, Improved mapping of Jupiter's auroral features to magnetospheric sources. J. Geophys. Res. Space Phys. **116**, 03220 (2011). https://doi.org/10.1029/2010JA016148

S. von Alfthan, D. Pokhotelov, Y. Kempf, S. Hoilijoki, I. Honkonen, A. Sandroos, M. Palmroth, Vlasiator: first global hybrid-Vlasov simulations of Earth's foreshock and magnetosheath. J. Atmos. Sol.-Terr. Phys. **120**, 24–35 (2014). https://doi.org/10.1016/j.jastp.2014.08.012

R. von Steiger, P. Shearer, T. Zurbuchen, Solar wind abundances from Ulysses-SWICS. AGU Fall Meeting Abstracts (2013)

R. von Steiger, N.A. Schwadron, L.A. Fisk, J. Geiss, G. Gloeckler, S. Hefti, B. Wilken, R.F. Wimmer-Schweingruber, T.H. Zurbuchen, Composition of quasi-stationary solar wind flows from Ulysses/solar wind ion composition spectrometer. J. Geophys. Res. **105**, 27217–27238 (2000). https://doi.org/10.1029/1999JA000358

J.H. Waite Jr., F. Bagenal, F. Seward, C. Na, G.R. Gladstone, T.E. Cravens, K.C. Hurley, J.T. Clarke, R. Elsner, S.A. Stern, ROSAT observations of the Jupiter aurora. J. Geophys. Res. **99**, 14 (1994). https://doi.org/10.1029/94JA01005

A.D.M. Walker, The Kelvin-Helmholtz instability in the low-latitude boundary layer. Planet. Space Sci. **29**, 1119–1133 (1981). https://doi.org/10.1016/0032-0633(81)90011-8

B.M. Walsh, T.A. Fritz, J. Chen, Simultaneous observations of the exterior cusp region. J. Atmos. Sol.-Terr. Phys. **87**, 47–55 (2012a). https://doi.org/10.1016/j.jastp.2011.08.011

B.M. Walsh, D.G. Sibeck, Y. Wang, D.H. Fairfield, Dawn-dusk asymmetries in the Earth's magnetosheath. J. Geophys. Res. Space Phys. **117**, 12211 (2012b). https://doi.org/10.1029/2012JA018240

B.M. Walsh, E.G. Thomas, K.-J. Hwang, J.B.H. Baker, J.M. Ruohoniemi, J.W. Bonnell, Dense plasma and Kelvin-Helmholtz waves at Earth's dayside magnetopause. J. Geophys. Res. Space Phys. **120**, 5560–5573 (2015). https://doi.org/10.1002/2015JA021014

B.M. Walsh, J. Niehof, M.R. Collier, D.T. Welling, D.G. Sibeck, F.S. Mozer, T.A. Fritz, K.D. Kuntz, Density variations in the Earth's magnetospheric cusps. J. Geophys. Res. Space Phys. **121**, 2131–2142 (2016a). https://doi.org/10.1002/2015JA022095

B.M. Walsh, M.R. Collier, K.D. Kuntz, F.S. Porter, D.G. Sibeck, S.L. Snowden, J.A. Carter, Y. Collado-Vega, H.K. Connor, T.E. Cravens, A.M. Read, S. Sembay, N.E. Thomas, Wide field-of-view soft X-ray imaging for solar wind-magnetosphere interactions. J. Geophys. Res. Space Phys. **121**, 3353–3361 (2016b). https://doi.org/10.1002/2016JA022348

G.K. Walters, Effect of oblique interplanetary magnetic field on shape and behavior of the magnetosphere. J. Geophys. Res. **69**, 1769–1783 (1964). https://doi.org/10.1029/JZ069i009p01769

C. Wang, H. Li, J.D. Richardson, J.R. Kan, Interplanetary shock characteristics and associated geosynchronous magnetic field variations estimated from sudden impulses observed on the ground. J. Geophys. Res. Space Phys. **115**, 09215 (2010). https://doi.org/10.1029/2009JA014833

Y. Wang, J. Raeder, C. Russell, Plasma depletion layer: its dependence on solar wind conditions and the Earth dipole tilt. Ann. Geophys. **22**, 4273–4290 (2004a). https://doi.org/10.5194/angeo-22-4273-2004

Y. Wang, J. Raeder, C. Russell, Plasma depletion layer: magnetosheath flow structure and forces. Ann. Geophys. **22**, 1001–1017 (2004b). https://doi.org/10.5194/angeo-22-1001-2004

Y. Wang, J. Raeder, C. Russell, Plasma depletion layer: the role of the slow mode waves. Ann. Geophys. **22**, 4259–4272 (2004c). https://doi.org/10.5194/angeo-22-4259-2004

Y.L. Wang, J. Raeder, C.T. Russell, T.D. Phan, M. Manapat, Plasma depletion layer: event studies with a global model. J. Geophys. Res. Space Phys. **108**, 1010 (2003). https://doi.org/10.1029/2002JA009281

Y. Wang, D.G. Sibeck, J. Merka, S.A. Boardsen, H. Karimabadi, T.B. Sipes, J. Šafránková, K. Jelínek, R. Lin, A new three-dimensional magnetopause model with a support vector regression machine and a large database of multiple spacecraft observations. J. Geophys. Res. Space Phys. **118**, 2173–2184 (2013). https://doi.org/10.1002/jgra.50226

Z. Wang, J. Cao, A.G. Michette, Depth-graded multilayer X-ray optics with broad angular response. Opt. Commun. **177**, 25–32 (2000). https://doi.org/10.1016/S0030-4018(00)00576-9

B.J. Wargelin, P. Beiersdorfer, G.V. Brown, EBIT charge-exchange measurements and astrophysical applications. Can. J. Phys. **86**, 151–169 (2008). https://doi.org/10.1139/P07-125

B.J. Wargelin, M. Markevitch, M. Juda, V. Kharchenko, R. Edgar, A. Dalgarno, Chandra observations of the "dark" Moon and geocoronal solar wind charge transfer. Astrophys. J. **607**, 596–610 (2004). https://doi.org/10.1086/383410

R. Wegmann, K. Dennerl, X-ray tomography of a cometary bow shock. Astron. Astrophys. **430**, 33–36 (2005). https://doi.org/10.1051/0004-6361:200400124

R. Wegmann, K. Dennerl, C.M. Lisse, The morphology of cometary X-ray emission. Astron. Astrophys. **428**, 647–661 (2004). https://doi.org/10.1051/0004-6361:20041008

M.C. Weisskopf, B. Brinkman, C. Canizares, G. Garmire, S. Murray, L.P. Van Speybroeck, An overview of the performance and scientific results from the Chandra X-ray observatory. Publ. Astron. Soc. Pac. **114**, 1–24 (2002). https://doi.org/10.1086/338108

I.C. Whittaker, S. Sembay, J.A. Carter, A.M. Read, S.E. Milan, M. Palmroth, Modeling the magnetospheric X-ray emission from solar wind charge exchange with verification from XMM-Newton observations. J. Geophys. Res. Space Phys. **121**, 4158–4179 (2016). https://doi.org/10.1002/2015JA022292

M. Wieser, S. Barabash, Y. Futaana, M. Holmström, A. Bhardwaj, R. Sridharan, M.B. Dhanya, A. Schaufelberger, P. Wurz, K. Asamura, First observation of a mini-magnetosphere above a lunar magnetic anomaly using energetic neutral atoms. Geophys. Res. Lett. **37**, 05103 (2010). https://doi.org/10.1029/2009GL041721

R. Willingale, J.F. Pearson, A. Martindale, C.H. Feldman, R. Fairbend, E. Schyns, S. Petit, J.P. Osborne, P.T. O'Brien, Aberrations in square pore micro-channel optics used for x-ray lobster eye telescopes, in *Space Telescopes and Instrumentation 2016: Ultraviolet to Gamma Ray*. Proc. SPIE., vol. 9905 (2016), p. 99051. https://doi.org/10.1117/12.2232946

G.R. Wilson, T.E. Moore, M. Collier, Low-energy neutral atoms observed near the Earth. J. Geophys. Res. Space Phys. **108**, 1142 (2003). https://doi.org/10.1029/2002JA009643

M. Wiltberger, R.E. Lopez, J.G. Lyon, Magnetopause erosion: a global view from MHD simulation. J. Geophys. Res. Space Phys. **108**, 1235 (2003). https://doi.org/10.1029/2002JA009564

M. Wiltberger, R.S. Weigel, W. Lotko, J.A. Fedder, Modeling seasonal variations of auroral particle precipitation in a global-scale magnetosphere-ionosphere simulation. J. Geophys. Res. Space Phys. **114**, 01204 (2009). https://doi.org/10.1029/2008JA013108

M. Wiltberger, V. Merkin, J.G. Lyon, S. Ohtani, High-resolution global magnetohydrodynamic simulation of bursty bulk flows. J. Geophys. Res. Space Phys. **120**, 4555–4566 (2015). https://doi.org/10.1002/2015JA021080

S. Wing, P.T. Newell, T.G. Onsager, Modeling the entry of magnetosheath electrons into the dayside ionosphere. J. Geophys. Res. **101**, 13155–13168 (1996). https://doi.org/10.1029/96JA00395

S. Wing, P.T. Newell, J.M. Ruohoniemi, Double cusp: model prediction and observational verification. J. Geophys. Res. **106**, 25571–25594 (2001). https://doi.org/10.1029/2000JA000402

S. Wing, J.R. Johnson, P.T. Newell, C.-I. Meng, Dawn-dusk asymmetries, ion spectra, and sources in the northward interplanetary magnetic field plasma sheet. J. Geophys. Res. Space Phys. **110**, 8205 (2005). https://doi.org/10.1029/2005JA011086

M. Witte, Kinetic parameters of interstellar neutral helium. Review of results obtained during one solar cycle with the Ulysses/GAS-instrument. Astron. Astrophys. **426**, 835–844 (2004). https://doi.org/10.1051/0004-6361:20035956

M. Witte, H. Rosenbauer, M. Banaszkiewicz, H. Fahr, The ULYSSES neutral gas experiment—determination of the velocity and temperature of the interstellar neutral helium. Adv. Space Res. **13**, 121–130 (1993). https://doi.org/10.1016/0273-1177(93)90401-V

R.S. Wolff, B.E. Goldstein, C.M. Yeates, The onset and development of Kelvin-Helmholtz instability at the Venus ionopause. J. Geophys. Res. **85**, 7697–7707 (1980). https://doi.org/10.1029/JA085iA13p07697

G.L. Wrenn, J.F.E. Johnson, A.J. Norris, M.F. Smith, GEOS-2 magnetopause encounters—low energy (< 500 eV) particle measurements. Adv. Space Res. **1**, 129–134 (1981). https://doi.org/10.1016/0273-1177(81)90096-X

Y.-H. Yang, J.K. Chao, A.V. Dmitriev, C.-H. Lin, D.M. Ober, Saturation of IMF B_z influence on the position of dayside magnetopause. J. Geophys. Res. Space Phys. **108**, 1104 (2003). https://doi.org/10.1029/2002JA009621

T.K. Yeoman, P.G. Hanlon, K.A. McWilliams, A statistical study of the location and motion of the HF radar cusp. Ann. Geophys. **20**, 275–280 (2002). https://doi.org/10.5194/angeo-20-275-2002

B. Zhang, O. Brambles, W. Lotko, W. Dunlap-Shohl, R. Smith, M. Wiltberger, J. Lyon, Predicting the location of polar cusp in the Lyon-Fedder-Mobarry global magnetosphere simulation. J. Geophys. Res. Space Phys. **118**, 6327–6337 (2013). https://doi.org/10.1002/jgra.50565

B. Zhang, O.J. Brambles, M. Wiltberger, W. Lotko, J.E. Ouellette, J.G. Lyon, How does mass loading impact local versus global control on dayside reconnection? Geophys. Res. Lett. **43**, 1837–1844 (2016). https://doi.org/10.1002/2016GL068005

H. Zhang, D.G. Sibeck, Q.-G. Zong, N. Omidi, D. Turner, L.B.N. Clausen, Spontaneous hot flow anomalies at quasi-parallel shocks: 1. Observations. J. Geophys. Res. Space Phys. **118**, 3357–3363 (2013). https://doi.org/10.1002/jgra.50376

H. Zhang, K.K. Khurana, M.G. Kivelson, V. Angelopoulos, W.X. Wan, L.B. Liu, Q.-G. Zong, Z.Y. Pu, Q.Q. Shi, W.L. Liu, Three-dimensional lunar wake reconstructed from ARTEMIS data. J. Geophys. Res. Space Phys. **119**, 5220–5243 (2014). https://doi.org/10.1002/2014JA020111

J. Zhang, W. Poomvises, I.G. Richardson, Sizes and relative geoeffectiveness of interplanetary coronal mass ejections and the preceding shock sheaths during intense storms in 1996–2005. Geophys. Res. Lett. **35**, 02109 (2008). https://doi.org/10.1029/2007GL032045

T.-L. Zhang, J.G. Luhmann, C.T. Russell, The solar cycle dependence of the location and shape of the Venus bow shock. J. Geophys. Res. **95**, 14961–14967 (1990). https://doi.org/10.1029/JA095iA09p14961

T.-L. Zhang, K. Schwingenschuh, C.T. Russell, J.G. Luhmann, Asymmetries in the location of the Venus and Mars bow shock. Geophys. Res. Lett. **18**, 127–129 (1991a). https://doi.org/10.1029/90GL02723

T.L. Zhang, K. Schwingenschuh, H. Lichtenegger, W. Riedler, C.T. Russell, Interplanetary magnetic field control of the Mars bow shock—evidence for Venuslike interaction. J. Geophys. Res. **96**, 11 (1991b). https://doi.org/10.1029/91JA01099

T.L. Zhang, K.K. Khurana, C.T. Russell, M.G. Kivelson, R. Nakamura, W. Baumjohann, On the Venus bow shock compressibility. Adv. Space Res. **33**, 1920–1923 (2004). https://doi.org/10.1016/j.asr.2003.05.038

X.-W. Zhou, C.T. Russell, G. Le, S.A. Fuselier, J.D. Scudder, The polar cusp location and its dependence on dipole tilt. Geophys. Res. Lett. **26**, 429–432 (1999). https://doi.org/10.1029/1998GL900312

X.W. Zhou, C.T. Russell, G. Le, S.A. Fuselier, J.D. Scudder, Solar wind control of the polar cusp at high altitude. J. Geophys. Res. **105**, 245–252 (2000). https://doi.org/10.1029/1999JA900412

H.C. Zhuang, C.T. Russell, An analytic treatment of the structure of the bow shock and magnetosheath. J. Geophys. Res. **86**, 2191–2205 (1981). https://doi.org/10.1029/JA086iA04p02191

J.H. Zoennchen, U. Nass, H.J. Fahr, Exospheric hydrogen density distributions for equinox and summer solstice observed with TWINS1/2 during solar minimum. Ann. Geophys. **31**, 513–527 (2013). https://doi.org/10.5194/angeo-31-513-2013

J.H. Zoennchen, J.J. Bailey, U. Nass, M. Gruntman, H.J. Fahr, J. Goldstein, The TWINS exospheric neutral H-density distribution under solar minimum conditions. Ann. Geophys **29**, 2211–2217 (2011). https://doi.org/10.5194/angeo-29-2211-2011

B.J. Zwan, R.A. Wolf, Depletion of solar wind plasma near a planetary boundary. J. Geophys. Res. **81**, 1636–1648 (1976). https://doi.org/10.1029/JA081i010p01636